LESSONS

from the

GREAT
GARDENERS

Matthew Biggs trained at the Royal Botanic Gardens, Kew, is a member of the Woody Plant Committee of the Royal Horticultural Society, and panel member on BBC Radio Four's *Gardeners' Question Time*. A well-known broadcaster, gardening writer, and personality, he has written several books including *The Complete Book of Vegetables: The Ultimate Guide to Growing, Cooking and Eating Vegetable*s.

The University of Chicago Press, Chicago 60637
© 2016 by Quid Publishing Limited
All rights reserved. Published 2016.
Printed in China

25 24 23 22 21 20 19 18 17 16   1 2 3 4 5

ISBN-13: 978-0-226-36948-8 (cloth)
ISBN-13: 978-0-226-36951-8 (e-book)

DOI: 10.7208/chicago/9780226369518.001.0001

Library of Congress Cataloging-in-Publication Data

Biggs, Matthew, author.
  Lessons from the great gardeners : forty gardening icons and what they teach us / Matthew Biggs.
    pages cm
  Includes bibliographical references and index.
  ISBN 978-0-226-36948-8 (cloth : alk. paper) — ISBN 978-0-226-36951-8 (e-book) 1. Gardeners—Biography. 2. Gardening—History. I. Title.
  SB451.B554 2016
  635.092'2—dc23
                    2015033806

LESSONS

from the

GREAT

GARDENERS

Forty Gardening Icons & What They Teach Us

MATTHEW BIGGS

THE UNIVERSITY OF CHICAGO PRESS

Chicago

Contents

Judas tree
Cercis siliquastrum

Frederick Stern
Page 106

Introduction

Welcome to the lives of forty great gardeners, past and present, whose creativity and practical experience have revolutionized and refined ideas on gardens and gardening. Among their number are artists who painted with plants, inveterate collectors, eccentrics, experimenters, billionaires, and monarchs who found gardens and gardening the finest way to express their creativity or status. All had concepts and designs; some gardenmakers built the gardens themselves, relishing the opportunity to dig and delve, while others employed gardeners and those with specialist skills. However their dreams were realized, their aim was the same: to perfect the art of the garden.

NATURAL SELECTION

When you consider how many gardens have been created around the world since time began, compiling a definitive list of forty iconic gardeners and their gardens is in fact an enormous challenge.

The first task was to make a list of outstanding gardens and their creators, the best of their kind, and to ensure that all major styles were represented. Many of the decisions were difficult and the selection is inevitably subjective. André le Nôtre represents landscape gardeners above Lancelot "Capability" Brown, for no other virtue than that when it comes to landscaping on a grand scale, there is only one Versailles. And how do you select from all of the great cottage gardens or Japanese gardens, other than by making a personal choice as to which is the best?

The creators of outstanding gardens that have made the final forty are talented, artistic, unconventional, freethinking, sometimes eccentric, and have revolutionary ideas that are ahead of their time. All have been in a position to realize their dream. You can appreciate a garden through the principles of planting and design, but it is worth remembering that a garden reflects the personality and character of its creator, so there are glimpses of that too. Because of their groundbreaking approaches, we can learn from their innovations and experiences and thereby improve the quality of our gardens at home.

Gardeners were also chosen for their impact on the gardening world. Some are held in high esteem, not just because of the garden they created but because their knowledge and ideas have been disseminated to a wider audience through publication in journals, newspapers, and, most notably, books. This ensures they are influential in both life and after death, when their writings are read by later generations. Two notable examples are William Robinson and Gertrude Jekyll. The photographic feature pages dotted throughout this book show just how widespread the influence of these forty gardeners has been. The internet, television, and greater freedom of transport have added to their accessibility, allowing many more people to see and study the gardens they made than would have been possible in generations past. Many of the great gardeners of our era are still creative forces. Visit their gardens while you can: you may be fortunate enough to meet them.

The work of many of the forty gardeners has been endorsed by an award from the Royal Horticultural Society: the Veitch Memorial Medal "may be awarded annually to persons of any nationality who have made an outstanding contribution to the advancement of the art, science, or practice of horticulture." The story of James Veitch is among those recorded in these pages.

GARDENERS THROUGH TIME

Gardening ideas and styles have changed over time, and history informs the present and future. There is evidence of previous eras—ponds, flower borders, formal designs, and statuary—in all of the featured gardens.

Early gardens were functional. Gardeners filled them with vegetables and medicinal plants to feed and heal. The most productive areas were in the rich soils next to great rivers, like the Fertile Crescent of Mesopotamia along the Tigris and

When creating his water garden at Giverny, Monet was inspired by the simple elegance and sense of balance of Japanese gardens, like those at Ryōan-ji (see page 9).

Euphrates and the Nile and its delta. Later on, once settlement and civilization had taken hold, an upper class emerged, their flower gardens tended by servants. Here, decoration replaced necessity, turning gardening into an art.

In ancient Egypt, flowers, fruit, and vegetables were grown together, rectangular ponds became a popular feature and trees, planted for shade, were often associated with different gods. In ancient Iraq, the Assyrians created great pleasure gardens lined by formal rows of trees. When their empire was destroyed by the Babylonians, King Nebuchadnezzar II was said to have created the Hanging Gardens of Babylon, one of the wonders of the ancient world. The Persians who followed grew fragrant flowers and fruit trees in formal gardens with ponds, fountains, and rhylls; the Romans adopted their ideas, adding topiary, statues, and flowers like cyclamen, iris, and lavender. In the 7th century, the Moors created Islamic gardens with high walls, fountains, glazed tiles, and mosaics, and trees for fruit and shade. In the late 15th century, the Italian Renaissance revived classical ideas and proportion and balance became important. Gardens were often laid out in a grid pattern, based around a main axis running

from the villa. These Italian gardens contained statues, grottoes, and even hidden water features. In England, William Kent and Charles Bridgeman mixed formal and informal designs, while Lancelot "Capability" Brown set out to improve, not rework, nature.

This history and its influences is the reason why the book is arranged in chronological order, by the gardeners' date of birth, rather than the style or era they represent. You will find all of the styles mentioned above reflected in this book. Arranging them in chronological order also shows how the development of other art forms affected gardens. Impressionist painters, particularly William Turner, influenced the plantings of Gertrude Jekyll, and she, like her collaborator Edwin Lutyens, was part of the Arts and Crafts movement (*c.*1860–1910). Claude Monet's paintings of waterlilies growing in the pond at the bottom of his garden, arguably his most famous, also confirm gardening's contribution to and inclusion as one of the arts.

MAKING CHANGES

Sometimes changes in garden style are by evolution, at other times by revolution. William Robinson's rage against the rigid formality of the ornamental gardening of his time and his subsequent ideas on "wild" gardening are currently reflected in the work of Piet Oudolf and the New Perennial movement, and they continue to evolve. Roberto Burle Marx's reaction to historic European gardens in Brazil was to combine his skill as an abstract artist with his passion for native tropical plants, creating a new style that reflected the modern country—a horticultural revolution. Ideas on gardens and gardening continue to develop.

The gardens and gardeners mentioned in the following pages are inspirational, their impact unquestionable. It is an ideal opportunity for us to look and learn from their foresight and experience, to understand more about the art of gardening and, finally, like them, to realize our personal dreams.

Botanical illustrations

Illustrations of specific plants or groups of plants widely associated with each iconic gardener are dotted throughout. It may be a plant that features prominently in their famous garden, which you can look out for when visiting, a plant that appeared in every garden they created because it was a particular favorite, something they selected from cultivation or collected in the wild, or one bearing their name. Whatever the reason, they are plants that will forever be synonymous with the gardener or their garden, like Margery Fish and herbaceous geraniums, and *Victoria* 'Longwood Hybrid', named after Longwood Gardens in Pennsylvania, masterminded by Pierre S. du Pont.

Caucasian cranesbill
Geranium ibericum

Lessons from the greats

These practical tips or design ideas may be direct quotes from the gardener themselves or observations made by the author from techniques they used in their gardens.

Photographic spreads

These illustrate gardens from around the world that have been the work of the great gardeners, showing their designs, plant associations, or specific collections.

Somai

1480–1525

Japan

Japanese cherry
Prunus serrulata

This ancient garden tree was introduced to Britain from Canton in 1822. It was the first Japanese cherry planted in European gardens.

The design of Ryōan-ji, a UNESCO World Heritage Site and one of the most famous Japanese gardens in the world, is attributed to the artist Somai, but given that there was little tradition of recording the construction of gardens it is difficult to be absolutely certain. What is certain is that this abstract rock garden is one of the most enigmatic works in all Japanese cultural design. Viewing the garden is a spiritual experience and open to individual interpretation; it provokes the questions, you provide the answers. By way of contrast, the landscaping and planting in the surrounding parkland is full of traditional horticulture with clipped evergreens, spring blossoms, and vibrant fall color.

Hosokawa Katsumoto built Ryōan-ji in 1450. Around two decades later it was burned down in the Onin War (1467–77) but was rebuilt by his son Hokosawa Masamoto. It is believed that the Rock Garden was built in 1499, at the same time as the hojo building. Tradition recognizes the architect of the garden to be the monochrome artist Somai (1480–1525), who worked in association with the Zen temple garden of Daisen-in. However, this is open to debate. It is known that professional laborers helped to build the garden, probably *sensui kawaramono*—"riverbank workers as gardeners"—aided by Zen monks. The names Hikojiro and Kotaro are chiselled on the back of one of the 15 rocks; there is a tradition of laborers marking their work, so perhaps it was them. Over the centuries, the atmosphere in the garden has changed. The original design is thought to have had a background of "borrowed" landscape, the views disappearing as the trees matured.

THE ROCK GARDEN

It is believed that the design of the garden is heavily influenced by the tea ceremony, in which honesty, understatement, and rusticity are held in high esteem. The two principles of honesty and understatement (*wabi*) associate well with Zen Buddhism, which absorbed the architecture of the tea ceremony into temple design, resulting in gardens such as Ryōan-ji. The minimalism of *wabi* allows imagination to fill the spaces. Most commentators agree that the gravel represents a void (emptiness also being associated with Zen Buddhism); this void is also represented in Japanese culture by the space between lines in a drawing, pauses in music, and the distances between actors in Noh theater. It is often said that, in all aspects of architecture and garden design, it is not the spaces you fill but those you leave that count, so each part of the garden has its "gallery space." This is evidenced in some of the succulent plantings by the artist Jacques Majorelle in the Jardin Majorelle in Marrakech (see page 111).

Measuring 98 x 33 feet, about the size of a tennis court, the simple Zen garden at Ryōan-ji, the Temple of the Peaceful Dragon, is formed from clay walls, white gravel on the ground that is raked every day, moss, and 15 carefully placed rocks in five clusters. According to Nitschke in *Le Jardin Japonais*, the garden is meant to be viewed when seated on the veranda but whatever angle it is viewed from (except overhead) only 14 stones can be seen. At any vantage point, one of the rocks is always hidden from the viewer. In Buddhism, the number 15 signifies wholeness or completeness; if you can see 15, you have already attained enlightenment through deep Zen meditation. Their alignment is seen to perfection when viewed from the veranda; the monks meditated here, gazing directly into the blankness of the garden.

All but one of the rocks seem to be flowing from left to right. There are many interpretations of this—the composition of Chinese characters for "heart;" a tiger crossing the sea with her cubs (referring to the Zen parable of *Tora-no-ko watashi*); or, more obviously, islands in an ocean, with the gravel resembling water. In 2002, researchers at Kyoto University found that, when viewed from a certain angle, the layout of the rocks subconsciously evoked the outline of a branching tree. It could be that it was meant to reflect the surrounding trees, such as Japanese maples, which provide a vibrant background of fall color, emphasizing the simplicity of the rocks and white gravel.

However, because the meaning is unknown, it is incumbent on the individual to discover the meaning for themselves.

The garden should provoke deep meditation, whatever its spiritual significance to the individual.

Japanese flag iris
Iris ensata

This clump-forming perennial has been developed for centuries in Japan, where different regions favored different flower shapes.

THE SURROUNDING PARKLAND

The surrounding gardens contain temples, shrines, a tea room, and several ponds. The largest is the still, reflecting Kyoyochi ("mirror-shaped") Pond, covered by vast rafts of waterlilies. In the shallows, *Nelumbo nucifera* (the sacred lotus) blooms, its waxy white and pink flowers protruding well above the surface, while *Iris ensata*, the Japanese flag iris, lines the water's edge. There are two islands. One, accessed by a simple arched stone bridge, leads to a shrine surrounded by cloud-pruned conifers. The pond, with its reflections of the sky, clouds and surrounding trees, the waterlilies, the marginal planting, and the bridge, inspired Monet's garden at Giverny (see page 54).

Around the pool, clipped evergreens hide the stems of densely planted conifers and deciduous trees, providing structure and fall color, which also cloaks the surrounding hillside. The occasional *Diospyros kaki*—Japanese persimmon—shows its orange-red to orange-yellow and plum-purple fall color, followed by orange tomatolike fruits. On the islands and scattered through the parkland are

> *The vacant space of the garden, like silence, absorbs the mind, frees it from petty detail, and serves as visual guide—a means of penetrating through the realm of "the multitudes."*
>
> —Will Petersen

Persimmon
Diospyros kaki

Prunus serrulata, Japanese cherries, with their delicate pink blossom in spring and rich yellow, orange, and red color in fall. Other areas feature densely packed clipped spheres and moss-carpeted ground under trees where the lower branches have been removed. One of the most notable plants is *Camellia japonica* 'Kochouwabisuke', the fish-tail camellia; associated with the tea ceremony, the plant is said to be the oldest in Japan.

SO, WHAT?

We leave the answer about the meaning of the rock garden to Will Petersen in his article "Stone Garden," published in *Evergreen review* vol.1 no.4 (1957). Like all great art, the garden is perhaps a "visual *koan*" (dialogue, question, or statement). It remains in the mind, and, if it can be likened to anything, rather than "islands in the sea," it is the mind. It does not matter, therefore, what materials the garden is composed of, what is important is the mind that interprets the essentials. The garden exists within ourselves; what we see in the rectangular enclosure is, in short, what we are.

SOMAI

Although the garden (see photograph overleaf) reflects on the culture and spiritual enlightenment relating to Zen Buddhism, the principles of proportion, illusion, and choice of materials can be adopted when creating gardens in other styles.

→ The Rock Garden slopes slightly to the left, allowing water to drain away, and the west wall slopes slightly to the south, creating the optical illusion of space, depth and perspective. Using illusion and perspective, as in the garden at Cloudehill (see page 202), is not only fun but creates an interesting sense of space.

→ The 6 foot wall around three sides of the rock garden is made from a mixture of earth and rapeseed oil, which will stand years of weathering and does not reflect the light from the white gravel. The inner face of the wall is 3 inches higher than the outer face, giving the wall greater stability. Building a mud or rammed earth wall in your garden is an interesting experiment that will be attractive and long-lasting.

→ All but one of the rocks seem to be flowing from left to right. When placing rocks, note their different faces; take these shapes into account when creating a rock garden. Two flat surfaces placed close together give the appearance of a rock twice the size, which has split; placing several together suggests the rock strata. Used in this way, the rocks will look more impressive and have a greater impact in the landscape.

Japanese maple
Acer palmatum

Many cultivars have been raised from this species. They form large bushes or small trees with attractive leaves and vivid red, orange or yellow fall color.

→ The design of the Rock Garden generates tension, drawing the viewer to contemplate the mystery of Zen. It can't be photographed in its entirety; the dimensions make that impossible. It is worth experimenting with dimensions and their impact in your own garden.

There is a strong association between gardens and religion. Christians believe God created a garden in Eden (the word "paradise" means garden). The first botanic gardens were collections of the work of the Creator. Islamic gardens are also full of religious symbolism. It is understandable, then, that the Zen Buddhist temple at Ryōan-ji is also surrounded by a garden, a peaceful, spiritual place, away from the bustle of the outside world. In Zen Buddhism, meditation is the path to enlightenment. "Zen" is the Japanese pronunciation of the Chinese word *Ch'an*, taken from the Sanskrit, meaning "meditation." Where better to declutter the mind than by focussing your attention on the empty space and carefully placed rocks in the Rock Garden?

Wang Xianchen

b. 1500s

China

Lady Banks' rose
Rosa banksiae

Named for Lady Banks, the wife of Sir Joseph Banks, one of the greatest directors of the Royal Botanic Gardens, Kew, England.

The Humble Administrator's Garden is one of the finest in the garden city of Suzhou and a UNESCO World Heritage Site. It marked the return to the simple life for Wang Xianchen, a civil servant who retired from work to create a garden. With its perfectly placed buildings, elegant architecture, and sense of balance and proportion, the garden is a landscape in miniature, which has been modified over the years. There is remarkable detail in the landscape and buildings; it is a garden where visitors can look and learn.

During the Ming Dynasty, the Imperial censor Wang Xianchen gave up his office and returned to his hometown, feeling he was unable to administer anything but a garden. He then spent 16 years building a house and developing its 13 acre gardens in the Ming style, though it has been altered over the years. Its name, the Humble Administrator's Garden, came from an essay by the Jin writer Pan Yue: "To cultivate my garden and sell my vegetable crop…is the policy of a humble man."

The garden, built on marshland, was completed in 1526 and is said to have cost a boatload of silver. Like gardens in many cultures, it was laid out as a series of pictures that reveal themselves as you walk: "awakening the reminiscence of the Venetian scenes, in the area south of the Lower Yangtze, [which] are archaic, rustic, extensive, and naturalistic…" (*An Introduction to Chinese Tourism*). Jean Denis Attiret, French court painter to the Qianlong Emperor, wrote: "One admires the art with which this irregularity is carried out. Everything is in good taste, and so well arranged, that there is not a single view from which all the beauty can be seen; you have to see it piece by piece." Chinese gardens are often referred to as living landscape paintings. Although they are surrounded by white walls, a pure backdrop for the art of the garden, many also "borrow" the landscape beyond, a principle known as *jie jing*. It is a prominent feature in the Humble Administrator's Garden, particularly when you look beyond to the Beisi Pagoda. This principle is encapsulated in a motto displayed in the garden museum: "Anything of value is to be taken in and the trite is to be ignored."

The garden is divided into three main sections: east (originally known as the country retreat), west, and central. Water, around which the whole garden is based, is a major feature of Chinese gardens, bringing energy while purifying and cooling the air. The softness of water balances the rigidity of stone—when two opposites, the *yin yang*, are in balance, it produces positive, rejuvenating *qi* energy. Here, the water reflects the details of the surroundings, creating balance, tranquillity, and the harmony of the garden.

Japanese aralia
Fatsia japonica

This fine architectural plant thrives in any type of well-drained soil, either in sun or part shade. Flowering in early fall, it is a valuable source of late nectar for insects.

PAVILIONS AND HALLS

It is traditional to have a large number of buildings in Chinese gardens and the Humble Administrator's Garden, with 48, has more than most, yet does not feel congested. They vary in size, are ornately crafted, in proportion, and, as always, they harmonize with the landscape. Several of the buildings have interconnecting galleries attached to them; these narrow covered corridors protect visitors from the sun and rain. One of the most elegant is the Small Flying Rainbow Bridge (Xiaofeihong), the only covered bridge in the garden. Within the central expanse of the lake, a series of interconnecting bridges and paths link the four main islands. The construction attempts to create scenery by piling up earth and rocks; each is beautifully landscaped and most are topped with pavilions. There are 18 in all, their names suggesting the views or sensory experiences on offer: for example, Distant View, Listening to the Rain, and "With whom shall I sit."

The Hall of the Distant Fragrance sits near a pond filled with *Nelumbo nucifera*, the pink-and-white-flowered sacred lotus; when they are in flower, the hall is filled with their fragrance. The lotus flower represents purity; they grow in mud but are cleansed as they emerge through the surface of the water. A carved screen of ginkgo wood divides the chief building in the western part of the garden into two halls. Near to the north hall, the Mandarin Ducks Hall is "a large pond with five colored waterlilies and 36 pairs of mandarin ducks and it looks as though it is richly ornamented with brocade," records the notebook of Zhen Shuai.

Beautifully weatherworn limestone rocks, used for edging and landscaping, are often softened with plants.

Rocks represent the structure of the world, their shape a reminder of how the gentle power of water gradually erodes rigidity. Granite is used mainly, though not exclusively, for courtyards, paths, and bridges. The garden is linked by intricate paths, winding through a landscape of shrubs, pines, slender bamboo, and evergreens. There are trees such as *Maclura tricuspidata*, the Chinese silkworm thorn, with red, mulberrylike fruit; *Ilex cornuta*, Chinese holly, with unusual rectangular leaves; azaleas, with blooms of bright pink and white; and evergreen *Fatsia japonica*, with thick, large-lobed leaves. Arches are laden with white-flowered wisteria and white double-flowered *Rosa banksiae* var. *banksiae*, and bright yellow *Jasminum mesnyi* pours over a wall. A shade house in the nursery is crammed with pink, white, and cerise azaleas; nearby is a display of over 700 penjing, or bonsai, in ornate ceramic pots on plinths and tables, all elegantly trained and some extremely ancient.

The whole garden is a refined mix of elegance, balance, and poise on a micro and macro scale.

Azalea
Rhododendron **species**

Rhododendrons are among the many flowering plants trained by bonsai enthusiasts, and specimens in bloom are a wonderful sight. Despite their small size, most bonsai are hardy and should be grown outdoors.

WANG XIANCHEN

In Chinese gardening, hard landscaping (like paths, temples, and bridges) regularly takes preeminence over plants. Many of the plants are trimmed, pruned, or chosen for their geometric shape. This discipline and control creates a sense of balance and calm.

→ Bamboos in borders are thinned so the canes are widely spaced and then underplanted with groundcover, creating a pleasing effect. The bamboo offcuts are used to make fences.

→ Buildings in the garden can be attractive, whatever their use. Even sheds can be painted and decorated with plants; this is especially useful in small gardens, where the shed is difficult to hide. You could add trellising, then a rose or honeysuckle, or windowboxes filled with bedding plants or lavender.

→ You may not have the space to replicate the rose-covered arch at the Humble Administrator's Garden or the long laburnum arch at Bodnant Garden (see page 98), but you can put a climber-covered arch over your garden gate or front door or use it to divide areas of your garden.

→ Meandering paths are not only a feature of the landscape and a means of access; they also unify the garden, tying it together. Make sure your paths are well maintained to avoid accidents and that formal paths are wide enough at right-angled junctions to enable you to turn a wheelbarrow comfortably. There should be a distinction between main paths—where two people can walk side by side—and access paths, for weeding and watering.

Chinese holly
Ilex cornuta

This plant's spined leaves, slow growth and compact formation make it ideal for restricted spaces. However, the large red berries are often few in number.

→ Moon gates and windows are used to frame views; achieve the same effect by cutting a "window" in trellising then covering it with climbers so you focus on the chosen view.

→ Azaleas in pots are grown outdoors then brought in when they are flowering. You can use the same technique for similar plants from the glasshouse or garden at home.

André Le Nôtre

1613–1700

France

Confederate (or star) jasmine
Trachelospermum jasminoides

This evergreen twining plant is
renowned for its beautiful flowers and
fragrance. It can be grown outdoors
against a sheltered, sunny wall.

André le Nôtre was the greatest landscape gardener of all time. He worked at a glorious
time for the aristocracy in France, when the wealthy lavished fortunes on their gardens
and estates, vying for preeminence in the art of horticulture. He created many gardens
that were outlandish extravagances, notably at Vaux le Vicomte, but they reached their
zenith in the landscape at Versailles. Despite his success on such a grand scale, he still
remained in touch with the soil and retained this interest until the end.

Born into a family of gardeners to King
Louis XIII, André Le Nôtre served his
apprenticeship at the Tuileries in Paris and
was allowed to inherit his father's post as
Master Gardener in 1637. He also studied
under the painter Simon Vouet, where he
learned about perspective and drawing, and
the architect François Mansart, who noted his
considerable talent.

His first great opportunity came in 1656,
when Nicolas Fouquet, Finance Minister to
Louis XIV, commissioned Le Nôtre, architect
Louis Le Vau and artist Charles Le Brun to
design Vaux le Vicomte. It was here that his
genius as a landscape architect became

apparent. No one had ever created a garden
like it in France; three villages were levelled
and the river diverted to make a canal.
He extended great blocks of trees from the
parterres, which gradually diminished,
accentuating the perspective and relating
the position of the trees to fountains and
statuary. He also obtained maximum
reflection by regulating water levels.

VERSAILLES

To celebrate the completion of his chateau
and gardens in 1661, Fouquet staged a
sumptuous party honoring the king, with

walks in the garden, supper, comedy, a ballet staged by Molière and Lully, and a huge firework display. But the king had been plotting. "At 6pm Fouquet was king of France, but by 2am he was nobody," wrote Voltaire. A month later Louis XIV had Fouquet imprisoned and, after confiscating his furniture and paintings and removing his trees and statuary, he gave the job of transforming Versailles, with its 23 square miles, to the three geniuses who created Vaux le Vicomte. With his vast experience as an architect and hydraulics engineer, duties as General Controller of Buildings, scholarly contacts at the Academy of Sciences, and with monks who were knowledgeable in optical geometry, André Le Nôtre was perfectly placed to succeed.

At Versailles, Le Nôtre refined his ideas: the principal walks were bisected by secondary walks around groves; trellises emphasized perspectives and formed vast walls of green; and he played with shadow and sunlight, alternating shady groves with open parterres. The parterres and principal walks were flanked by innumerable statues (scale models of the sculptures were positioned before they were carved) and clipped yew hedges. Versailles became a center of excellence for the topiarist's art and imagination.

His main vistas were awe-inspiring; it was gardening for the ego, a reminder of Louis's grandeur. Le Nôtre created *L'allée Royale* (the Royal Alley), grottos, and a grand canal like no other, while tile workers from Flanders, marble cutters from the Pyrenees, Italy and Greece, masons,

sculptors, and metalworkers lent their skill to the construction of the glorious fountains. It was a monumental task. Earth was transported in wheelbarrows, trees were conveyed by horse and cart from all the provinces of France, and thousands of men—sometimes whole regiments—were employed to realize Le Nôtre's magnificent design. Such was the scale that there were 350 gardeners on the payroll, more than any other craftsmen.

Lemon
Citrus × limon

Renowned for its fruits and sweetly fragrant flowers, the lemon is one of the most popular conservatory plants. Leave the mature fruit on the plant and pick them as needed.

ANDRÉ LE NÔTRE

Le Nôtre's ability to think on a grand scale yet still be concerned with the finer details is a trait well worth developing. It's like putting together a jigsaw puzzle: see the bigger picture first, then you can work out where the smaller pieces should be placed.

→ Le Nôtre was the first to give so much prominence to sculptures and topiary as garden decoration. Topiary is the perfect medium to bring structure to the garden, whether spheres, spirals, or cubes. Mature specimens can be expensive to buy, but you can grow, trim, and create your own. Where the disease "box blight" is a problem, use alternative plants such as *Ilex crenata* 'Dark Green,' *Phillyrea angustifolia f. rosmarinifolia* or *Hebe topiaria.*

→ At Versailles, Le Nôtre invented a new type of parterre: the water parterre. Two large rectangular pools reflecting the sunlight replaced turf and raked gravel. Le Nôtre regarded light as part of the decoration of a garden, like foliage. This can be achieved on a smaller scale at home.

→ The 17th-century gardens gave trees renewed importance. With Le Nôtre, the progression of their height and density reinforced the effects. There are 15 groves at Versailles, all densely planted. Where space is limited, the effect of several trees can be achieved by planting multi-stemmed rather than single-stemmed trees.

Yew
Taxus baccata

Yew grows more rapidly than is often thought. Its dark, dense foliage makes it ideal for creating garden "rooms." With good drainage it will grow in almost any type of soil.

→ At Versailles there are 27 miles of trellises. Trellising can be used to create semi-permeable barriers, playing on visitors' curiosity to tempt them further into the garden; it is ideal where there is little ground space and can be a feature in its own right, not just utilitarian. Trellises can be created to your own design and spacing using roof battens, treated with timber preservative before they are installed. When putting trellis on a wall, put hinges along the horizontal base and attach them to battens for extra depth. Hinges allow the trellis to be removed from the wall for pointing, painting or pruning. Remember, too, that a square can also be a diamond.

→ Le Vau built a new orangery at Versailles and there are now 900 orange trees in pots. Citrus, particularly lemons, are ideal for fruit and fragrance. They should be kept above freezing (indoors) over winter and brought onto the patio when the danger of frost has passed.

The message was that not only was the king in control of France, he was even in control of the landscape. With an unlimited budget, it might seem quite an easy thing to do, but Le Nôtre's eye for proportion was exceptional.

—Diarmuid Gavin

THE ESSENCE OF GREATNESS

What set Le Nôtre apart was his innate sense of scale, balance, and proportion, always taking account of the topography, and his attention to visual effects such as the circular basin on the Tuileries parterre, which was offset from the center to correct the optical distortions. He dominated the landscape with a strong main axis across a shallow valley so he could play with changes in level, for which he devised architectural features. He had the ability to imagine, plan, and design on an extraordinary scale.

His sweeping proposals were not always well received by the clients. Mlle de Montpensier at Choisy refused to fell trees to open up the view and Louis XIV's brother, whose garden was on several levels in a situation described by William Robinson as "one of the most beautiful that gardening man could desire," failed to implement all of his plans.

But it was not all about grand designs. The most original creation of Le Nôtre's later years was the *Bosquet des sources*, a network of tiny irregular streams, small enough to step over, surrounding about 20 turf islands, each just large enough for a gaming table and chairs.

Drawings by Le Nôtre are extremely rare; most were quick sketches or comments on a plan. He worked closely with relatives,

Perennial ryegrass
Lolium perenne

Ryegrasses are often included in grass mixes for domestic lawns, because they are robust enough to withstand wear and tear.

who interpreted his ideas, entrusting his assistants with the land surveying and the final drafting of his projects.

Despite his status and success, he remained a practical gardener, retaining responsibility for the upkeep of the parterres and surrounding *allées* at the Tuileries and the espaliered jasmine along the terrace. He also stocked the flower beds and kept the gardens between the parterres and terrace "full of flowers, particularly in winter."

He had a house at Versailles and a home near the Tuileries; he lived out his life surrounded by his art collection, including paintings by Poussin and Lorrain, sculptures, tapestries, bronzes, and porcelain. In 1693, with no heir (his children died in infancy), he gave the king the best works. He was kind, spontaneous, witty, and amiable. His work was appreciated; he was respected for his integrity and worked for all his clients as diligently as he did for the king.

Louis XIV rewarded him with honor and they remained friends for over 40 years.

▶ This view, from the terrace above the Orangerie at Versailles, shows permanently planted topiary and orange trees in Versailles planters (which, themselves, are a classic design that remains popular today). The orange trees are taken indoors during winter and placed outdoors each spring. Their placement also emphasizes the structure and formality of the garden, with perfectly straight lines and symmetry echoing the architecture of the buildings. Water is often a feature in French gardens, the reflections adding to the sense of symmetry, order, and tranquility.

▲ Box and yew topiary have decorated the gardens at Versailles since the 17th century, inspiring the costumes worn for parties. On February 25, 1745, during a particularly icy spell, the wedding of the dauphin was celebrated by a masked ball named *Le Bal des Ifs* (The Yew Tree Ball). The bridegroom was transformed into a gardener and his wife, a flower seller. The art of trimming topiary has been passed down through the generations; the gardeners at Versailles maintain nearly seven hundred topiary hedges and trees trimmed into sixty different shapes. These forms emphasize the effects of perspective and are an essential part of the structure of the formal garden.

Philip Miller

1691–1771

United Kingdom

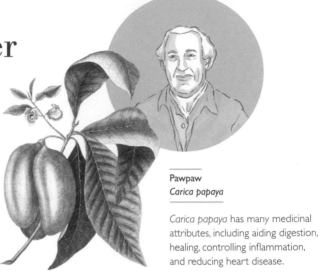

Pawpaw
Carica papaya

Carica papaya has many medicinal attributes, including aiding digestion, healing, controlling inflammation, and reducing heart disease.

Gardener, botanist, and writer Philip Miller was the most distinguished and influential British gardener of the 18th century. Under his curatorship, the Chelsea Physic Garden in London flourished. Through his many correspondents he was the first to grow hundreds of new introductions that came into Britain. The details of their cultivation and his other horticultural experiences were published in *The Gardeners Dictionary*, which reached its eighth edition in his lifetime. He was consulted by aristocrats and presidents and was held in such high esteem that he was known to many as *Hortulanorum princeps* (Prince of Gardeners).

Little is known of Miller's early life; he is believed to have been born in Deptford or Greenwich. His father ran a market garden in Deptford, southeast London, and Philip became a "florist and ornamental Shrub nurseryman" in St George's Fields, South-wark. When the Worshipful Society of Apothecaries needed a new gardener for their Physic Garden at Chelsea, he was recommended to Sir Hans Sloane, the garden's benefactor, as one "to go forward, with a curiosity and genius superior to most of his profession" (Patrick Blair, *Botanik*

Essays, 1720); Miller was appointed in 1722. Part of Sloane's deed of covenant was that they present annually to the Royal Society "fifty specimens or samples of distinct plants well dried and preserved and which grew in the said garden, together with their respective names…until the compleat number of two thousand have been delivered." No plant was to be offered twice, demanding a continuous flow of new introductions to the garden.

Miller's diligent correspondence with other gardeners and collectors helped his cause, leading to the introduction of a large number of rare seeds and plants from around the world. He was a major contributor to the doubling of the number of species in cultivation between 1731 and 1768, mainly from the Cape of Good Hope, the East Indies, and North America.

John Bartram, a farmer and plant collector in North America, was particularly helpful. Forty-eight of his American introductions were first cultivated in England at the Physic Garden. He wrote: "Mr. Miller… has made his great abilities well known by his works as well as his skill in every part of gardening and his success in raising seeds produced by a large correspondence." Miller then shared his knowledge in his *Gardeners Dictionary*. His expertise led to the cultivation of melons, pineapple, and pawpaw in beds of fermenting oak bark at the Chelsea Physic Garden. Sloane was even able to present some pineapples to the king. (Johanna Lausen-Higgins, an authority on the history of pineapple cultivation in Britain, suggests it may have been King George II since Miller had mastered pineapple growing by the 1730s.) Miller transformed the garden; it became one of the best known in Europe, and his reputation burgeoned.

A "COMPLEAT BODY OF GARDENING"

Miller became a household name after the publication of *The Gardeners Dictionary*, which covered horticulture, arboriculture, agriculture, and wine making. Eight editions (1732–68) were printed in his lifetime. It was translated into German, Dutch, and French and described as "the most compleat body of

gardening extant." He also produced an *Abridgement* in eight editions and a cheaper *Gardeners Kalendar* in 15 editions. Such was the range of formats and prices that everyone, from small-scale gardeners to aristocrats, could afford a copy. Even George Washington, the first president of the United States, referred to the *Dictionary* and *Kalendar* when redesigning the landscape at Mount Vernon.

The Gardeners Dictionary was presented in a dictionary format, with each entry providing the common name, a description, and cultivation details. For strawberries,

Cucumber tree
Magnolia acuminata

This vigorous plant soon becomes a large, spreading tree. Its common name describes the shape of the immature fruits.

Miller writes: "Kinds must have a greater share of room according to their different degrees of growth, as for example the scarlet strawberry, should be planted a foot square plant from plant and the Hautboy sixteen or eighteen inches difference each way." There are recommendations for pruning, watering, propagation, including grafting and layering, and planting; and advice on edging, groves, hedges, hotbeds, and "troublesome vermin."

GARDENING CONSULTANT PAR EXCELLENCE

Pehr Kalm (1716–99), a Finnish pupil of Carl Linnaeus (inventor of the binomial nomenclature system), said of Miller that "the principal people in the land set a particular value on this man." Between 1740 and 1753, he was paid 20 guineas a year by the 4th Duke of Bedford to visit his estate at Woburn Abbey, Bedfordshire, to inspect the hothouses and gardens and to prune his trees. Among the duke's most prized possessions were his plantations of North American trees, among them *Pinus rigida*, the pitch pine, and *Abies balsamea*, the North American fir.

The Duke of Richmond at Goodwood, West Sussex, sought Miller's advice, and he also curated the collection of North American plants grown by Lord Petre at Thorndon Hall, Essex. (Petre inherited the estate in his teens and immediately began filling it with American introductions on a large scale: he had 900 *Liriodendron tulipifera*, tulip trees, alone.) These connections led to the Royal Society consulting Miller on agricultural matters, including grape varieties for South Carolina. Miller often advised growers in France and Italy, receiving seeds from both countries, and was a member of the Botanic Academy at Florence.

ALL IN A NAME

Miller used the scientific names proposed by J.P. de Tournefort's *Institutiones* (1700) and did not use the modern binomial naming system until the eighth edition of *The Gardeners Dictionary* (1768). In it he proposed several hundred new botanical names for species that Linnaeus had not recorded, making the edition of particular importance to botanists.

Miller was distinguished for his vast theoretical knowledge of plants and "especially by his skill in their cultivation" (John Rogers, *The Vegetable Cultivator*, 1839). He trained Sir Joseph Banks, the friend of George III who developed the Royal Botanic Gardens at Kew; William Aiton, who created a small physic garden at Kew; and William Forsyth of *Forsythia* fame. In 1763 his son Charles became the first curator of Cambridge Botanic Garden.

John Rogers proclaimed Miller "a benefactor to mankind: medicine, botany, agriculture, and manufactures are all indebted to him." His knowledge of living plants was unsurpassed in breadth in his lifetime and, as his contemporary Peter Collinson noted, he "raised the reputation of the Chelsea Garden so much that it excels all the gardens of Europe for its amazing variety of plants of all orders and classes and from all climates…"

PHILIP MILLER

Gardening knowledge has grown significantly since Miller's day, but the basic principles remain the same. Much of his advice is relevant today, with some minor adaptations. Below are some extracts taken from the eighth edition of his *Gardeners Diary* (1768).

→ Main crop beetroot: "This sort requires a deep light soil, for as their roots run deep in the ground, so in shallow ground they will be short and stringy… the roots will be fit for use in the [fall] and continue good all the winter." Choose your varieties accordingly.

→ Thyme: "May be propagated by slips [cuttings] or seeds may be sown in spring. They delight in undunged ground, where they will propagate themselves by their trailing stalks and require no other care but to keep them free from weeds."

→ Rhubarb: "In [fall], the leaves of these plants decay, then the ground should be made clean and in the spring, before the plants begin to put up their new leaves…the ground should be hoed and made clean again."

→ Eggplants: "They must be propagated by seeds which must be sown in a moderate hotbed in March and when the plants come up, they must be transplanted into another hotbed, about four inches asunder observing to water and shade them until they have taken root."

→ Cistus: "May be intermixed with other shrubs where they will make a pretty diversity; and in such places where they are sheltered by other plants they endure the cold much better than where they are scattered singly in borders."

→ Earwigs: "These are very troublesome vermin in a garden…others hang the hollow claws of crabs and lobsters upon sticks in diverse parts of the garden, into which these vermin get." A more traditional take is to use an upturned flower pot filled with hay.

Black Jamaica pine
Ananas comosus

"Like to a cone of the Pine tree, which we call a pineapple for the forme… being so sweete in smell… tasting… as if Wine, Rose water and Sugar were mixed together." (John Parkinson, *Theatrum Botanicum*, 1640)

Thomas Jefferson

1743–1826

United States

Edible pea
Pisum sativum

Peas are one of many crops that taste better when eaten fresh, immediately after harvesting. As Jefferson noted, they are best home grown.

Better known as the third President of the United States and principal author of the Declaration of Independence, Thomas Jefferson was also a passionate and knowledgeable gardener, who began noting the plants growing in the garden and woodlands around his home as a child. When Jefferson was 14, his father died, leaving him around 2990 acres of land and 36 slaves. Thirteen years later, he began building Monticello (Italian for "Little Mountain") on an exposed mountain top near Charlottesville, Virginia.

Jefferson designed the imposing neoclassical house, workshops, and garden at Monticello, the latter in the landscape style of the 19th century. The garden is divided into several sections: the flower, vegetable and fruit gardens, the West Lawn, and the surrounding grounds. There were also 20 oval beds around the house, containing 105 different herbaceous plants, laid out in 1807. The oval-shaped West Lawn, surrounded by a sinuous flower border, was laid out in 1808. This informality reflected Jefferson's interest in the picturesque style, which he had admired during his visit to England in 1786.

By 1812, he had organized the border into 3.3 yard sections, each numbered and planted with a different flower, the majority supplied by Bernard McMahon (commemorated in the genus *Mahonia*), a nurseryman and author of *The American Gardener's Calendar*, which Jefferson often used as a practical guide. Many plants that he grew, such as roses and sweet williams, were well established in Europe, and bulbs including hyacinths and tulips (the most commonly mentioned bulb in his later *Garden Book*) were widely planted alongside novelties such as the winter cherry (*Physalis alkekengi*).

Exotics and experimentation

Although he received up to 700 different kinds of seed annually from the Jardin des Plantes in Paris, Jefferson was passionate about the diverse native flora. A quarter of the plants growing at Monticello were of North American origin, including *Jeffersonia diphylla* (the "twinleaf," named for Thomas Jefferson in 1792) and plants such as *Fritillaria pudica*, the yellow fritillary. The latter was discovered during the Lewis and Clark Expedition (1804–06) across the western part of North America, commissioned by Jefferson (Lewis and Clark are also commemorated in the plant genera *Lewisia* and *Clarkia*). Jefferson also trialled "exotics" such as French figs, Mexican peppers and English peas, alongside beans and salsify collected by Lewis and Clark. "I am curious to select one or two of the best species or variety of every garden vegetable, and to reject all others from the garden to avoid the dangers of mixing or degeneracy," he wrote.

Vegetables were initially cultivated at Monticello along a 330 yard-long slope; it was not until 1806 that terracing was cut out of the mountain, supported by a stone wall that in some parts is 12 feet tall. At its midpoint is a garden pavilion with a Chinese railing and pyramidal roof. The vegetable garden, covering 2 acres, contained 24 square plots, planted according to which part of the plant was being harvested—leaves, roots, or fruits. It contained a wide range of common crops, plus orach, nasturtiums, endive, corn salad, sesame for salad oil,

Sea kale
Crambe maritima

French artichokes and new introductions such as tomatoes and broccoli, and prized sea kale (*Crambe maritima*) for its blanched young shoots in spring.

In addition, Jefferson's 8 acre "fruitery" included the South Orchard, which formed a horseshoe encompassing two vineyards and "berry squares" of raspberries, currants, and gooseberries. There were also "submural beds" by a sunny wall, which created a microclimate for figs and strawberries. Here he grew 150 cultivars of 31 kinds of temperate fruit, among them quince, almond, nectarine, and pear. Not all were successful, and some

Fig
Ficus carica

It can be grown as a small tree, in containers or trained as a fan against a sheltered, sunny wall. The fruit are delicious eaten straight from the plant, having been warmed by the sun.

Sugar maple
Acer saccharum

Jefferson loved trees and had particular favorites (or "pet trees"), including mulberry, peach, and Sugar Maple. Some had their crown raised to expose the beauty of their bark.

of the Mediterranean crops in particular struggled with the climate. However, peaches flourished: Jefferson wrote to his granddaughter in 1815 that "we abound in the luxury of the peach." Apples and cherries were prolific, too, and it is said that his collection of fruit trees represented the finest cultivars available to an early 19th-century gardener.

EARLY PEAS

Notes in his *Garden Book*, which he kept between 1766 and 1824, reveal Jefferson's meticulous approach to gardening. Here he would note his observations, successes, failures ("killed by frost Oct 23") and precise cultivation details. He wrote of asparagus peas: "$^2/_3$ pint sow a large square, rows $2^1/_2$ feet apart and 1 f. and 18 I. apart

in the row, one half at each distance." The book also reveals what was clearly Jefferson's favorite vegetable: peas.

There are references to one particular sheltered border that was used to grow 15 varieties of early peas. By using different varieties and successional sowing, Jefferson was able to harvest peas for two months, from mid-May to mid-July. He took this so seriously that every spring he held a competition with his friends to grow the first English peas of the year. The winner hosted a celebratory dinner in which the peas were served. The usual winner was not Jefferson himself but a neighbor, George Divers. As Jefferson's grandson recalled: "A wealthy neighbor [Divers], without children, and fond of horticulture, generally triumphed. Mr. Jefferson on one occasion had them first, and when his family reminded him that it was his right to invite the company, he replied, 'No, say nothing about it, it will be more agreeable to our friend to think that he never fails.'"

During the years following Jefferson's death, the garden at Monticello fell into decay. However, in 1939 the Garden Club of Virginia launched a two-year restoration program and visitors can now see the gardens as they would have looked during Jefferson's own lifetime.

No occupation is so delightful to me as the culture of the earth, and no culture comparable to that of the garden…I am still devoted to the garden. But though an old man, I am but a young gardener.

—Thomas Jefferson

THOMAS JEFFERSON

Thomas Jefferson's *Garden Book* is a meticulous record of activities and events in his garden, with information about the flowers, fruit, trees, and vegetables he planted, including varieties' sowing places and dates and harvesting time, and comments on the weather. From this the modern gardener can gain numerous insights.

→ Make sowings of vegetables, annual flowers, or bulbs such as gladioli over a period of several weeks. Known as "successional sowing," this will extend the harvesting or flowering period.

→ Take advantage of protected microclimates, such as those created in a sheltered corner or against a sunny wall, by growing early crops or tender fruit such as grapes or figs.

→ When designing a garden, remember that a curved pathway slows progress, encouraging visitors to stop and look at what is growing around them.

→ Raise the "crown" of woody plants and bamboos by removing lower branches or leaves to reveal the beauty and color of the bark.

→ With careful selection, native plant species are more suited to the local fauna and environment, less prone to pests and diseases, and generally more robust.

→ Most importantly, learn from Jefferson's example by keeping a garden book or diary. This is not just for historical reference—it provides valuable information about your garden and the plants growing in it, so you can make informed decisions in the future. For instance, recording the varieties and cultivar names of vegetables you have planted enables you to focus on those that perform consistently well in your garden.

Peach
Prunus persica

Peaches can be pollinated by hand using an artist's paintbrush to transfer pollen from one flower to another, ensuring a good crop of these delicious, succulent fruit.

Most vegetables are short-term crops that grow rapidly and are harvested before reaching maturity. Jefferson's kitchen garden is in an open sunny site, with good air movement, reducing the chance of fungal diseases and pests, which prefer still conditions, and aiding wind-pollinated crops like sweet corn. A sunny slope is perfect; the soil warms up faster in spring, making it ideal for early crops, cold air rolls down the slope, reducing the impact of late frosts, and sunshine encourages insects for pollination. Ideally, vegetables should be planted so the sun shines on both sides of the rows as it moves through the sky during the day.

Sir Joseph Paxton

1803–1865

United Kingdom

Dwarf Cavendish banana
Musa cavendishii

This species of banana, named for William Cavendish, 6th Duke of Devonshire, was later mass produced on plantations around the world.

Joseph Paxton, born in Milton Bryan, Bedfordshire, came from humble beginnings. He was the youngest son of nine children; his father, William, a farm worker, died three months before Joseph's 17th birthday, plunging the family into poverty. Yet this didn't hold him back from success: he went on to become an influential and famous horticulturist, publisher, writer, landscaper, glasshouse designer, and engineer, and "the busiest man in England," as Charles Dickens described him. Perhaps best known for designing the Crystal Palace for the Great Exhibition in 1851, he was highly influential in the development of glasshouse design and has many a gardening achievement to his credit, not least of all being first to cultivate and bring to flower the enormous *Victoria regia* (giant Amazon waterlily).

Paxton's gardening career began at Battlesden Park, Bedfordshire, where he received a thorough education in horticulture. By November 1823 he had gained a position at the Horticultural Society's garden at Chiswick. Keen and industrious, he educated himself in the library and "was training the creepers and newly introduced plants on the walls." It was here he met the 6th Duke of Devonshire, a wealthy and passionate plant collector, who,

impressed by the 23-year-old Paxton's enthusiasm and ability, invited him to be superintendent of his gardens at Chatsworth House, one of the grandest estates in Britain.

Paxton certainly didn't waste any time getting started. Within a short time of starting, he had made significant changes, with "vegetables where there had been none, fruit in perfection, and flowers" appearing. Paths were improved, pipes were replaced, gardens were laid out after building works on the north wing of the house and the orangery was repaired. He also introduced rules for the gardeners and encouraged them to keep diaries for their education.

GLASSHOUSES

In 1828 he turned his attention to the kitchen garden and also experimented with glass structures to house exotic plants, which the Duke loved. Paxton preferred wood to the fashionable iron, insisting it was cheaper and easier to maintain, while still producing excellent fruit and plants for the house. One example is the conservative wall (see photograph page 41), so called because it conserves heat, a series of eleven separate units built against a wall in 1838, with a system of flues, hot water pipes, and canvas curtains to protect tender plants in winter. In 1848 he added the wood and glass frame. There are figs, nectarines, peaches, apricots, and tender shrub camellias growing against the wall; two *Camellia reticulata* 'Captain Rawes' planted in 1850 survived until 2000 and 2002. The most famous of the glass-houses, the Great Conservatory, covering 0.75 acres, was completed in 1840. It was a revolutionary design of curving lines and glass set in a ridge and furrow pattern, which improved stability and optimized light. People came from all over Europe to marvel at the "tropical scene with a glass sky." Inside were ponds full of aquatic plants, ferns, mosses, and brilliantly colored and lush-leaved plants, including *Musa cavendishii*, the dwarf Cavendish banana. Paxton planted it in 1835 and the following year harvested 100 fruit.

Around this time, "the striking butterfly Orchid (*Oncidium papilo*) so impressed the Duke of Devonshire that he began to assemble a collection of orchids." They were gathered from all over the world and became the largest collection in England.

Paxton set about experimenting with orchids, concentrating on perfecting methods of cultivation—he was delighted to find that the orchids thrived in the loamy, highly fibrous soil from the woods at Chatsworth. He advocated separate houses for plants from different climates. What is now the Vinery, built c.1834, remains the sole survivor of three orchid glasshouses Paxton constructed. Several orchids were named after Chatsworth, Paxton or the Duke: for example, *Coelogyne cristata* 'Chatsworth', introduced from India in

Giant Amazonian waterlily
Victoria amazonica (regia)

Victoria amazonica is still a major visitor attraction in gardens. It was first discovered in 1801. Its leaves grow extremely rapidly, up to five square feet per day.

SIR JOSEPH PAXTON

Joseph Paxton was renowned for his natural talent, thirst for knowledge, willingness to experiment, and appetite for hard work. These qualities ensured his success and are a source of inspiration for any ambitious gardener. He also met the right people.

→ While studying at Chiswick, Paxton used the library extensively to further his understanding. The ultimate aim is to know what you are doing and why you are doing it. This reduces the element of chance and allows you to experiment. All gardeners should strive to expand their knowledge, whether at an educational course, by joining a specialist society, or simply by consulting the wealth of gardening literature.

→ Paxton's first great landscaping work was laying out and planting a pinetum. Conifers make fine garden trees, providing there is sufficient space. Their shape and form is their virtue, so they are better grown as specimen trees, rather than being grouped together, where this attribute is lost. Where conditions allow, deciduous conifers such as *Metasequoia glyptostroboides*, the dawn redwood, are certainly worth considering for their fresh green spring growth, elegant form, russet fall color and winter silhouette.

→ When experimenting with orchids, Paxton found that plants responded to different conditions, depending on their habitat. Many gardeners complain that they cannot grow orchids or houseplants; the reason is that they have yet to find the right habitat in their house. Research the background to plants before you buy, aim to match the plant to the location, and don't be afraid to move them elsewhere if they aren't thriving. If they die, don't take it personally—work out why and try again. Successful gardening is based on practical experience.

Cedar of Lebanon
Cedrus libani

This majestic tree, found in Turkey, Syria and Lebanon, is now classed as vulnerable in the wild, making those growing in cultivation even more valuable.

1835 by a young Chatsworth gardener, John Gibson, *Dendrobium paxtonii*, and *Stanhopea devoniensis*.

THE GREENEST OF BOUGHS

Alongside his passion for glasshouses, Paxton found time to undertake many great landscaping works. In 1830–31 he created a pinetum—one of the first in the country, at a time when extensive botanical collections and fine gardens were regarded as status symbols among the wealthy. He planted *Cedrus libani* (cedar of Lebanon); *Pseudotsuga menziesii* (Douglas fir), the seeds of which he transported from London wrapped in his hat; *Araucaria heterophylla* (Norfolk Island pine), which arrived from Ireland; and *Pinus parviflora* (Japanese white pine). Paxton later oversaw the planting of an arboretum, a systematic succession of trees in accordance with botanical classification, and became successful in moving mature trees, the heaviest weighing 8 tons.

THE CROWNING GLORY

In 1849, William Hooker, the Director of the Royal Botanic Gardens at Kew, informed Paxton that some seeds of *Victoria regia* (now *V. amazonica*), the giant Amazon waterlily, had germinated. Kew had received the seeds from the Amazon, but as yet no one had succeeded in getting the plant to flower. Paxton requested a seedling and built a water tank. There followed a race between

Butterfly orchid
Oncidium papilo

Paxton and Hooker to produce the first flower. Paxton collected the plant in August and by early October it boasted a leaf 4 feet across. The plant thrived and in November a large bud opened to a flower of pure white, with a pineapplelike fragrance. The first bud was cut, which faded almost immediately, but it was revived by Paxton, who put fine sand near its base "to imitate the activity of nibbling fish." Paxton sped it to Queen Victoria at Windsor and Hooker graciously conceded defeat.

Paxton, who was knighted in 1851, remained head gardener at Chatsworth until 1858. He produced a number of publications, and in 1841 he co-founded the horticultural periodical *The Gardeners' Chronicle* with John Lindley, Charles Wentworth Dilke and William Bradbury, and later became its editor.

Paxton died at Sydenham, Kent, in 1865 and is buried at Chatsworth; a peasant gardener, who rose to mix with kings.

Botany—the science of the vegetable kingdom, is one of the most attractive, most useful, and most extensive departments of human knowledge. It is, above every other, the science of beauty.

—Sir Joseph Paxton

 Bedding schemes like this one in the Cottage Garden at Chatsworth are regularly changed, so head gardeners need to be organized and plan ahead, as did their predecessor, Paxton. Understanding the requirements of plants is the key to good displays. Tulips with forget-me-nots is a tried and tested combination for spring. Forget-me-nots often suffer from powdery mildew in hot dry weather, so plenty of compost is dug into the soil to retain moisture. Spent flowers are removed from tulips as they fade by gathering all of the petals in one hand before they fall.

The conservative wall (so called because it conserves heat) was one of Paxton's early experiments in frost protection and ripening. It was built after industrialization reduced the price of glass. The wall was originally part of a protected walkway to the stables but Paxton realized that, being south-facing, it could be used to increase exotic fruit production for the house. Similar, temporary frames are used by gardeners to protect peach and apricot blossom early in the season—usually a wooden frame covered in clear polythene, with gaps at the sides allowing access for pollinating insects. Constructed of several sections, they are light and easy to move and although they are not as grand or permanent as Paxton's structures, they are still extremely effective.

James Veitch, Jr.

1815–1869

United Kingdom

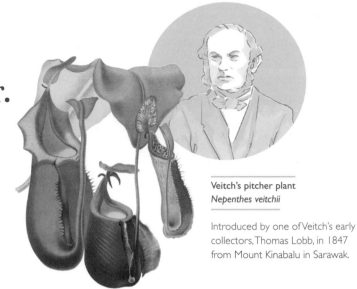

Veitch's pitcher plant
Nepenthes veitchii

Introduced by one of Veitch's early collectors, Thomas Lobb, in 1847 from Mount Kinabalu in Sarawak.

James Veitch, Jr. was born in Exeter, Devon. He came from a long line of nurserymen, the famous Veitch family. He grew up to be an accomplished businessman and had a huge impact on the growth of the family business. He also received many accolades as an exhibitor and was involved with setting up the RHS Plant Committees. Today, the Veitch name is perhaps most associated with the Veitch Memorial Medal, awarded annually in recognition of outstanding contributions to the advancement of the art, science, or practice of horticulture.

For over a hundred years and five generations, the Veitch family ran the largest and most influential nurseries in Europe. After John Veitch established the first nursery in Killerton, Devon, he and his son James bought land in Mount Radford, Exeter, and built up the business, later adding nurseries in prestigious locations in London. The Veitch family were shrewd businessmen and passionate plantspeople, renowned for employing their own plant collectors to traverse the globe to inhospitable climes in the hunt for plants. There were 22 in all, including members of the family and Ernest Henry Wilson from Chipping Campden in Gloucestershire, regarded as one of the greatest collectors. These collections were the basis of their success, and they provided top-quality plants for the gardener. In their era, the acquisition of new and desirable plants was a prestigious affair; many were planted with great ceremony on country estates, becoming status symbols, notably *Araucaria araucana* (then *Araucaria imbricata*), the monkey puzzle, and what is now *Sequoiadendron giganteum*, the giant redwood, both introduced by their first collector, William Lobb.

A DYNASTY MAINTAINED

It was into this prestigious gardening heritage that James Veitch, Jr. was born. At the age of 18 he was sent to London for two years to gain experience, spending a year at Alfred Chandler's nursery in Vauxhall (a specialist in camellias) and a second with William Rollison of Tooting. Veitch, Jr. had been passionate about orchids since he was a child. In lieu of a fee for training his son, James Veitch Senior agreed to buy some of their orchids; this formed the basis of the collection for which Veitch's became famous.

After his time in London, Veitch, Jr. returned to Exeter and devoted his energies to the gradual extension and improvement of Mount Radford, making it eventually "one of the finest nurseries of the day" for the wide range of rare and desirable plants it sold. He moved to the Royal Exotic Nursery in Chelsea in 1853, starting the business with his father: "Here he was brought into contact with all the leading horticulturists; and his inestimable personal qualities, his sound sense, and his energetic manner, soon lifted him into a very influential position in the gardening world which he maintained for many years." (James H. Veitch, *Hortus Veitchii*.) The following year, he relinquished his associations with the Exeter nursery, and from then on the Royal Exotic

Giant redwood
Sequoiadendron giganteum

Nursery was run separately. He also acquired large sites on the outskirts of London. On the death of his father in 1863, Veitch, Jr. became head of James Veitch and Sons.

A great nurseryman, "he occupied a no less worthy position as a cultivator and exhibitor. Indeed, Mr. Veitch was a thorough cultivator;" he entered with "great spirit the growth of dahlias for competitions." He grew his own plants for showing, of the highest quality, "so abundant in quantity and arranged with so much skill and taste" that it was rare for Veitch to be "either absent from or occupying a secondary place." All the accolades he received were good publicity for the Veitch nursery.

RHS PLANT COMMITTEES

From 1856 to 1864, Veitch, Jr. was a member of the council of the RHS. He came up with the notion of forming the Fruit and Floral Committees, an idea "which was first broached and talked over, even until the small hours in the parlour at the Royal Exotic Nursery" (*Hortus Veitchii*).

There are now seven specialist RHS committees, among them the Woody Plant, Bulb and Herbaceous Plant committees, each with a maximum of 24 members.

They help the RHS and its members in horticultural development and plantsmanship, bringing gardening in all its forms to a wider audience. They also improve standards of plant breeding, cultivation, and propagation. On a practical level, they review Award of Garden Merit (AGM) plants and assess plants that are proposed for the award. To receive an AGM, plants are first trialled, usually at Wisley or one of the other RHS gardens, and their performance is assessed by a forum of experts. Plants that are awarded the AGM receive it "for their consistency, reliability, and excellence for ordinary garden use." Committee members also write for RHS publications and participate at their events, giving talks and demonstrations. All are positive contributions, encouraging the enjoyment of gardening for all.

Veitch, Jr. died of a heart attack at the age of 54 at his home, Stanley House, King's Road, Chelsea, on September 10, 1869, and is buried in Brompton Cemetery. "This was no ordinary man. Zeal and energy pervaded his every action… and a warmth of friendly feeling that can be adequately gauged only by those who knew him." (*Hortus Veitchii.*)

As for the Veitch family fortunes: by the onset of the First World War they had introduced an extraordinary 1,281 plants, among them 498 glasshouse plants, 232 orchids, and 153 deciduous trees, shrubs, and climbers, which had a massive impact on the great gardens and conservatories of Britain.

Cattleya
Cattleya × exoniensis

John Dominy, a plant breeder at Veitch's nursery, flowered this plant, the first hybrid between two genera, *Laelia crispa* and *Cattleya mossiae*, in 1863.

THE VEITCH MEMORIAL MEDAL

In October 1896, there was a desire to perpetuate the memory of James Veitch, Jr. A request for ideas was published in *The Gardeners' Chronicle*; there were many suggestions and the proposal of William Thomson, Head Gardener of Dalkeith Palace, was accepted. He suggested a Veitch Medal to be awarded "to the man who during the year shall make the most important additions to our garden productions, whether by importation or hybridization." First struck in 1873 in gold, silver and bronze, most early medals were awarded for exhibits; 20 years later, additional awards were given to individuals who distinguished themselves in horticulture.

In 1922, Sir Harry Veitch, the Principal Trustee, arranged that the RHS should award the medal in future. It is the only RHS medal that can be awarded to people who are not British subjects. So far, over 100 of almost 500 medallists have come from outside the UK. Medallists include Ernest Henry Wilson (1906), George Forrest (1927), Francis Kingdon-Ward (1934), Vita Sackville-West (1955), Sir Harold Miller (1962), Joy Larkcom (1993) and Piet Oudolf (2002).

JAMES VEITCH, JR.

James Veitch, Jr. was a scientist and a showman with a passion for plants. The plants produced in the Veitch nursery for sale and for the show bench were of equal quality. There are no records of suggestions for showing from James Veitch, Jr. himself, but the following should be considered.

→ Always read the show guide to find out exactly what is required in each class, then follow that instruction to the letter.

→ Complete your entry forms early, in case there are queries.

→ Make sure you have plenty of top-quality plants to choose from, to ensure you have exactly what you need on the day. This often involves producing many specimens so you can choose those that are in top condition.

→ Visit the show you plan to enter the year beforehand, to check the quality of the opposition and chat with them if possible. Show guides often have basic growing information for novices, so anyone can have a go—don't be afraid to ask.

→ Keep a diary containing sowing or planting dates, cultivation details; anything that will be useful the following year.

→ Keep a constant lookout for pests and diseases. Use barriers such as horticultural fleece: prevention is better than cure. Do not enter anything that is less than perfect; judges are on the lookout for imperfections.

→ If flowers haven't opened due to inclement weather, they can be encouraged to do so by directing a gentle, warm stream of air from a hairdryer over them.

→ Be prepared for "growing for showing" to become a time-consuming but enjoyable obsession. There is no success without sacrifice.

Mysore clock vine
Thunbergia mysorensis

This climber for a warm, frost-free climate is widely grown as a garden plant and has received the Award of Garden Merit from the RHS.

Thomas Hanbury

1832–1907

United Kingdom

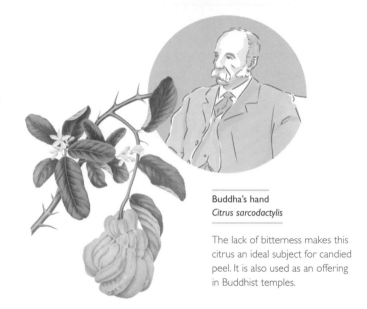

Buddha's hand
Citrus sarcodactylis

The lack of bitterness makes this citrus an ideal subject for candied peel. It is also used as an offering in Buddhist temples.

Born in Clapham, Surrey, Thomas Hanbury went on to become an extremely successful businessman, primarily in China. His portfolio included exporting tea and silk to Europe and trading in currency for Rothschild's. When the American Civil War interrupted cotton exports, he bought up Chinese cotton to supply to Britain and invested his capital, becoming the largest property owner in Shanghai at that time. Working with his brother Daniel, a pharmacologist, he created an acclimatization garden at La Mortola, a villa on the Italian Riviera, based on economic and medicinal plants. It gained a worldwide reputation for the range and diversity of the collection. He also donated Wisley garden to the RHS.

Thomas first saw La Mortola on March 25, 1867, on a boat trip along the Mediterranean coast. Within weeks he had bought the first part of what is now a 45 acres estate and retired, living in La Mortola until his death. Thomas was ably assisted in the creation of the gardens at La Mortola by his brother Daniel (1825–75), a noted pharmacologist and botanist. Daniel was also a partner in Allen, Hanburys and Barry, pharmaceutical manufacturers, at a time when crude plant extracts were the source of most medicines. Thomas was more interested in aesthetics, while Daniel preferred the scientific, educational, and medicinal aspects of gardening. The historical garden that Thomas created had a good botanical collection and a strong horticultural tradition, and appealed to both brothers' tastes. Thomas also wanted to research the subject of acclimatization. At the time, it was thought that subtropical species might adapt to cooler European conditions if they were acclimatized in Mediterranean climes beforehand.

Planting begins at La Mortola

Over the years, local peasants collecting firewood and grazing livestock had almost denuded the land. The Hanbury brothers sowed seeds of native evergreens such as *Rhamnus alaternus* (Italian buckthorn), *Quercus ilex* (holm oak), and ivy along the boundaries, restoring the native scrub. Daniel also added *Cistus* species, the first three dozen brought from his father's garden in Clapham, London, the remainder as seeds from nearby populations. He also planted native *Pinus pinaster*, the maritime pine, along the shoreline.

The rough hillside sloping down to the sea was lined with Italian cypress and olive terraces. Most were retained and interplanted with exotic species. Thomas added avenues of Italian cypress, repaired the impressive

500 foot pergola, and furnished the garden with vases imported from China and antiquities found on the estate during excavations.

Thomas and Daniel planted in great numbers, with early plants coming from nearby nurseries. In fall 1857, Thomas's notes mention "Passion Flowers, Geraniums, Peonies, Cedars of Lebanon, and roses," and on November 5, 1868, a striking Mexican *Dahlia imperialis*, described as "very fine and making a great show." There were 80 *Acacia* species, a host of medicinal plants including *Aloe ferox* (the Cape aloe), *Catha edulis*, or khat, the leaves of which are a stimulant, and exotic fruit such as pawpaw and guava. At one time he grew every known citrus cultivar, though he was unsuccessful with crops such as coffee and tea. Daniel was particularly interested in succulents; in June 1868 he noted 40 species, the number doubling later that same year. It rapidly became a plant collection of distinction.

Until December 1868, Thomas and Daniel worked the garden themselves, with a little help from local laborers, but later they employed some 30 gardeners, usually overseen by a resident German botanist because of Daniel's many contacts with German institutions. The most notable, Ludwig Winter, worked there for six years and was principal architect of the garden after Daniel's work in the early years.

Needle palm
Yucca flaccida

E. A. Bowles notes in his book *My Garden in Summer* that this is "one of the best in the family for flowering generously." The flower spikes reach up to 4 feet tall.

A GARDEN GROWN FROM SEED

Most of the plants were raised from seed and, as the collection increased, there was a vast annual exchange. In an act of considerable generosity, seeds were sent out free of charge. In 1900, 6,378 packets were distributed; by 1908 it was 13,085. Printed on the back were Thomas's requirements. Seeds also arrived at the gardens in a global seed exchange, many from countries with a Mediterranean climate.

Thomas regularly sent lists to *The Gardeners' Chronicle* of plants "in flower in the open air" in January. Those blooming on January 10, 1874 numbered 103, including *Datura arborea* (angel's trumpet) and *Polygala myrtifolia* (the September bush); on January 10, 2015, there were 204. The tradition changed for a time, listing plants in flower on New Year's Day; there were 294 in 1895, 405 in 1898 and 1926 and 232 in 1985, long after Thomas had died. In 1889, Thomas published *Hortus Mortolensis*, cataloging plants growing in the garden; the second edition, published in 1897, contained 3,600 species, proving just how successful his garden had become.

Cape aloe
Aloe ferox

THOMAS HANBURY AND RHS GARDEN, WISLEY

In 1902, George Fergusson Wilson—a former RHS Treasurer—died, and the Council of the RHS immediately inquired if his property at Wisley was for sale. They had been searching for a new location to replace their current garden at Chiswick; however, their offer was rejected. Meanwhile, Sir Thomas made a secret offer to the Council; he would buy Wisley and give it to the society, on condition that a committee was established to manage it. Two weeks later, terms were agreed and in October 1902 the RHS announced the purchase of its new garden. In 1903, Sir Thomas presented the Wisley estate in trust to the society.

The original garden was the creation of George Fergusson Wilson. He purchased the site in 1878 and established the "Oakwood experimental garden" with the idea of making "difficult plants grow successfully." The garden acquired a reputation for its collections of lilies, gentians, Japanese irises, primulas, and water plants, and despite many changes it remains true to the original concept.

A garden of exotic plants, which in point of richness and interest has no rival among the principal collections of living plants in the world.

—Sir Joseph Hooker, retired director of the Royal Botanic Gardens, Kew, describing La Mortola in 1893

THOMAS HANBURY

Thomas Hanbury created the La Mortola garden to experiment with the theory that subtropical species might adapt to cooler European conditions, if they were acclimatized in Mediterranean climes beforehand. The contrast between these subtropical and European climatic types, however, was too great.

→ For most gardeners in cool temperate climates, "hardening off" is the gradual process of acclimatizing plants that have been grown indoors for two to three weeks, until they are robust enough to live outdoors. Bedding plants, ornamental or vegetable seedlings propagated under glass in spring, and tender or half-hardy plants that have been given winter protection all undergo this process. This usually involves putting plants outdoors in a sheltered spot during the day and bringing them indoors at night for two to three weeks, making sure the final planting date is after the danger of frost has passed.

→ Many thousands of plants were grown from seed at La Mortola. It is the ideal way of exchanging plants over long distances and does not damage the plant in its original habitat. Not all plants come true from seed (that is, are an exact copy of their parent), nor is it instant; but it is cheaper and, when "difficult" seeds germinate successfully, immensely satisfying. It is still, as Thomas Hanbury found, a good way of gaining access to unusual plants.

→ At La Mortola, the greatest threat is drought, so plants from other Mediterranean climates are planted out in fall, taking advantage of any winter rain, and the more resilient, subtropical plants in spring.

Mimosa
Acacia dealbata

Renowned for its bright yellow flowers and attractive foliage, mimosa needs a warm, frost-free location to flourish.

→ Thomas was eager to share the excitement of having plants flowering in his garden in January. Providing winter does not mean permanent snow, winter-flowering plants are possible. Among them are forms of daphne, wintersweet, jasmine, and snowdrop. Place them by a door or where you can view them from a window.

William Robinson

1838–1935

United Kingdom

Adam's needle
Yucca gloriosa

This medium-sized evergreen shrub is valued for its architectural form and flowers. Site this carefully, away from paths, as the leaves have spiny tips.

William Robinson was born in July 1838. His first job as garden-boy was at Curraghmore, Co. Waterford, "being that of carrying water from the…river…to the glasshouses." After several years of gardening in Ireland and London, he decided to devote himself to the study of "our Great Gardens and to the Literature of Horticulture." His travels in Europe and North America inspired a plethora of publications, formed his robust opinions on gardening, and shaped his vision of its future. Robinson's journal, *The Garden* (1871), and two influential best-sellers, *The Wild Garden* (1870) and *English Flower Garden* (1883), cemented his fame and continued influence, while his own garden at Gravetye Manor in Sussex, England reflected his ideas.

Robinson, a "pugnacious paradox" (Richard Bisgrove, garden historian), was also a gardener with a conscience. He was a follower of the artist, critic and "thinker" John Ruskin, who railed against the injustice of the Industrial Revolution that took children from their homes, forced them to work in factories, and treated them as expendable. Robinson made a similar analogy with the Victorian formal "bedding" popular at the time, which took plants from around the world, forced them in hothouses, planted them by the thousand, laid them out in disciplined lines, stripped them of their identity, and disposed of them once their job was done. To him, it was an evil that it was his duty to overthrow, to be replaced with a democratic style of gardening, free of discipline and control, where people and plants expressed themselves and their personalities freely, like the plants he had seen in nature and cottage gardens—the "wild garden." He was also aware of the link between art and gardening, where lines and colors flowed. "The gardener must follow the true artist…in his respect for things as they are, delight in natural form and beauty of flower and tree, if we are to be free from barren geometry, and if our gardens are to be true pictures." Gertrude Jekyll (see page 60) reacted against rigidity by painting with plants; Robinson, by mimicking plants in nature.

He promulgated his ideas by writing, publishing prolifically in books and journals. *The Garden*, first published on November 25, 1871, included Gertrude Jekyll and William Morris among its contributors. His ideas were not always new, but they were considered, developed, then refined. In a series of books published around 1870, he expounded on a range of subjects from subtropical to alpine gardening, explaining how plants and plantings could be incorporated into gardens.

In 1870, Robinson wrote in his book *The Wild Garden*: "My object is… to show how we may, without losing the better features of the mixed bedding… [naturalize or make] wild innumerable beautiful natives of many regions of the earth in our woods, wild and semiwild places, rougher parts of pleasure grounds,

etc, and in unoccupied places in almost every kind of garden." *The English Flower Garden* (1883), described as "the most widely read and influential gardening book ever written," went into 15 editions in his lifetime.

Robinson was a visionary. The first issue of *The Garden* (1871) contained an article on roof gardens, which he had seen in America. He believed that every roof in London should be flat, with a garden so that people could grow their own food. He proposed radical ideas about horticulture and society, believing that with correct government policies and a system that grew vegetables as intensively as he had seen in France, more people, particularly the poor, could be fed. He even wrote on asparagus cultivation, intending for it to be less of a luxury.

Everything he proposed was to benefit the nation, particularly the poor, through parks, green spaces, and fresh food. Around 40 years before the term "ecology" was

Summer snowflake
Leucojum aestivum

One form, known as 'Gravetye Giant', selected by William Robinson, reaches 35 inches tall and bears up to 8 slightly fragrant flowers. Robinson naturalized this under trees at Gravetye Manor.

WILLIAM ROBINSON

Robinson disliked the formal gardens popular in his time and encouraged gardeners to mimic nature. Most modern-day gardens have an element of "naturalistic style."

→ Robinson believed you should only grow plants that were hardy in your own country. This included natives and exotics from around the world, from similar climates and habitats, that could thrive outdoors.

→ After observing plants on his travels, Robinson became aware of a huge range of habitats, noting the plants that flourished in them. If this correlation between plant and habitat was translated into the garden by putting "the right plant in the right place," both plant and gardener would be happy.

→ Garden plants should not be subjected to the will of the gardener by growing them

unnaturally, but the gardener should consider their form and habit, and then conceive ways of using them inventively.

→ After visiting the Alps, Robinson wrote *Alpine Flowers for English Gardens*, suggesting for the first time that Alpines could be cultivated by creating rock gardens. They can also be displayed on a smaller scale in pots, troughs and crevice

gardens, the latter made using vertical rock, slate, or tiles sandwiched together with plants growing in the gaps.

→ Robinson advocated the use of groundcover plants between taller specimens "that expose no bare soil." It is also an attractive way of suppressing weeds and preventing soil erosion.

→ Robinson created the mixed border, predominantly herbaceous plants with some shrubs, a style that can be adapted to even the smallest garden.

→ Robinson's book *The Wild Garden* and the garden at Gravetye Manor demonstrated his ideas of natural plantings using drifts of daffodils or herbaceous plants, which are managed, but the intervention of the gardener goes unnoticed. By choosing the correct plants, this too can be achieved on a smaller scale.

Geranium rose
Rosa moyesii

Known widely for the hybrid 'Geranium', its geranium red flowers are followed by attractive fall color and large bright red, flagon-shaped fruits in winter.

coined, Robinson was already interested in the environment and sustainability, and was concerned about the use of fossil fuel, pollution, and climate change due to the felling of trees (he planted thousands on his estate). He was also aware of the importance of green spaces for public health, regarding parks as an indicator of civilization and believing in the mutual coexistence of nature and man.

GRAVETYE MANOR

Robinson's publications and shrewd business sense made him a wealthy man. He moved into Gravetye Manor, West Sussex, in August 1884 and "de-Victorianized" it, removing the rock gardens and shrubberies. With the help of Ernest George, one of the great architects of the day, he built the terrace and summer house. He laid out the garden and experimented with his "naturalistic approach."

Robinson wanted to understand plant communities better by enhancing the natural beauty of the gardens and woods. Even when old and in a wheelchair, he would scatter bulbs and seeds from a bag on his lap. Areas leading up to the house were a mix of formal and semiformal; beyond the house, meadows, coppicing, and woodlands were embellished with drifts of bluebells, aconites, and narcissi. Together with his influential friends, he urged and achieved the transformation of English gardening from stifling formality to the relaxed "natural" style that is familiar today.

Wood anemone
Anemone nemorosa

A blue form of this flower was named for William Robinson: *Anemone nemorosa* 'Robinsoniana'. He recorded in 1883 that he found it in Oxford Botanic Gardens, where it had been sent from Ireland.

ROBINSON'S LASTING LEGACY

Robinson talked about grasses long before the advent of the New Perennial movement (see page 196), and his ideas on gardening link with the New Wild Garden, the German school of garden design, and prairie gardens. Robinson believed that even in a small plot there was room for a wild space and his influence on growing your own, naturalistic planting, and understanding the benefits to health of green spaces mark him as a very forward-thinking man and a great and influential gardener.

He changed the Face of England. Grand Old Man of the New Gardening.

—*London Evening News*

Claude Monet

1840–1926

France

Foxglove
Digitalis purpurea

One year Monet reported that spring flowering was a complete failure. To avoid this, plants like foxglove are sown early indoors, then transplanted outside.

Claude Monet, the most famous Impressionist artist, created one of his greatest works by painting the earth with flowers. To do this, he embraced the art of gardening and became an enthusiastic and knowledgeable gardener. He loved flowers. If he was not painting he was planting, tending, or thinking about his garden; it became a place to experiment with color, light, and form before applying brush to canvas. His "most beautiful masterpiece," it inspired some of his most famous paintings, and is also a quintessential example of the garden as art.

Monet signed the lease for his house in Giverny on May 3, 1883. He was to transform what was a cider orchard and *potager* (an ornamental kitchen garden) on poorly drained alkaline soil above clay into a unique and inspirational garden.

He began by growing vegetables to feed the family: "My salon was the barn, all of us worked in the garden, I dug, planted, weeded and hoed, myself; in the evenings the children watered." Soon, the garden vied with the easel for his attention. It is divided by a *grande allée* (a wide avenue) sloping from the front door down to the road, with box-edged borders on either side. Over time, some plants were removed, others remained. Out went the box edging; apple trees were replaced by Japanese cherries and apricots and cypresses by metal arches, which still remain today. Several spruces stayed, but they were roughly pruned as a support for climbing roses. Apart from the circular beds by the house, the layout was based on rectangles, forming a grid of flower beds. The rough grass was transformed into a rectangular lawn splashed with clumps of bulbs, such as daffodils, and herbaceous plants, including peonies.

Paintbox planting

In 1886, Monet visited the Dutch bulb fields, an experience that changed his use of structure and color. On his return he began planting single beds with one kind of plant in bold blocks of color. He continued by creating 38 smaller "paintbox beds" planted in pairs running down the garden, each reflecting the season in annuals and biennials.

Monet used perspective and light in the garden as he would in his paintings. He planted deep saturated colors in the foreground, with paler varieties behind, making borders seem longer. He planted blue flowers under trees, enhancing the color of the shadow cast by their branches, while bright oranges and reds were planted where they were backlit and glowing in the setting sun.

Climbers such as roses and clematis draped over frames, some of which he designed himself. Beds and borders burgeoned through the seasons, starting with *Hesperis matronalis* (sweet rocket) and tulips in spring, followed by magnificent lilies, roses, and other brightly colored annuals and perennials. He also had a passion for

bicolored flowers, such as bearded iris, and used color to contrast with the pink and what is sometimes known as "Monet's green" of his house. Edges of borders were softened with pinks, saxifrage, aubretia and, most notably, the nasturtiums either side of the *grand allée*; often by the end of the season these tides of green flowed to the center, forming a bejewelled green carpet.

"I must have flowers always and always," he said, so the whole garden merges seamlessly from spring to fall. Monet liked the simplicity of single wild flowers, which were allowed to self-seed; yellow-flowered *Verbascum thapsus*, the great mullein; foxglove, willow and poppies—every plant was vetted and selected to suit his needs.

Water forget-me-not
Myosotis scorpioides

Forget-me-nots featured prominently at Giverny. Other types, such as the water forget-me-not pictured, are also good in gardens.

Common tulip
Tulipa gesneriana L.

Carl Linnaeus, the father of modern taxonomy, gave this name to a group of old tulip cultivars in 1753.

The water garden

In later years, Monet said, "As my financial situation improved, I expanded until one day I crossed the road and started the water garden." First he dug a small pond (later enlarged), around which he planted aquilegias, roses, lupins, and *Gunnera*, creating structure and texture and softening the edge. Inspired by his Japanese prints, he planted willows, acers, and bamboos and wisteria over the Japanese bridge, all mirrored in the water. On his visit to the bulb fields in Holland, Monet had observed how the heaps of bright flowers (removed so all of the energy goes into the bulb, not into making unwanted seeds) were reflected in the water, where they were piled by the canals for collection: "On these little canals we see spots of yellow, like colored rafts in the blue reflection of the sky."

Waterlilies sat serenely on the surface, tropical waterlilies in vibrant tones—amethyst blue, deep plum and fiery sunset blends of yellow, pink, and apricot—and hardy varieties in coral pink, white, yellow, orange, red, and burgundy; all perfectly placed, allowing space for reflections.

Monet claimed he was "good for nothing except painting and gardening" and was very

It was in summer that one should have seen him, in this famous garden that was his luxury and his glory, and on which he lavished extravagances like a king on his mistress.

—Louis Gillet

"hands on." He planted, he dug, he could be found with "arms black with compost." In later years, when he could afford gardeners, he was still immersed in the garden's welfare. Before one of his painting trips he left detailed instructions for his head gardener. "Sow approximately 300 pots of poppies, 60 pots of sweet peas…Start blue sage and blue waterlily in a pot (greenhouse). Plant dahlias and water iris. From the 15th to the 25th start the dahlias into growth. Take cuttings from the shoots before my return; think about the lily bulbs."

The consummate gardener, Monet attended flower shows, leafed through catalogs, visited specialist suppliers and bought plants by mail order. (His waterlilies came from the world-renowned Latour-Marliac nursery, founded in 1875 and still in existence. His orders are still in their archive. Monet was the first to paint what was then a botanical novelty in Europe—non-white waterlilies.) He chatted about plants with his gardening friends, and is said to have lifted potatoes with Renoir and exchanged plants and seeds (his Japanese agent brought plants from Japan). He was exceedingly generous with gifts. Anyone who expressed an interest received plant material; parcels from his garden were sent by train all over France. "All my money goes into my garden," he said, and, "I am in raptures."

He loved color, he loved life, and he "loved all flowers," from simple wild flowers to garden favorites and rarities, in equal measure. There are artist gardeners; and then there is Monet.

CLAUDE MONET

A visit to Giverny is a must for any gardener. Who would miss a chance to walk through a three-dimensional, living Impressionist painting by Claude Monet? As you walk through the garden, squint, or look through your camera's viewfinder with it slightly out of focus, or if you are short-sighted, stop and remove your glasses; this will create the effect of looking at an Impressionist painting.

→ Monet was inspired by the Japanese prints he collected. Find your inspiration, something that chimes with your own ideas. It may be nature, a painting, or a pattern on china. Gardeners find inspiration everywhere.

→ Monet's garden is densely planted; this not only creates the cottage garden effect but covers the ground, and reduces the chance of weeds germinating.

→ The wisteria over the bridge, which is carefully pruned and trained, is a wonderful example of "less is more." Although wisterias are sometimes planted to cover the front of a house, they have a strong impact when only a few stems are trained and look attractive, though more formal, as espaliers.

→ Monet wanted a constant display of flowers. This can be achieved by using bulbs, annuals, and perennials and collecting specific groups of colors. He grew many blue plants, a color not widely represented in the world of flowers.

Siberian flag iris
Iris sibirica

These elegant early summer flowering plants with slender leaves and stems thrive in moist soil in sun but tolerate shade. There is a wide selection of hybrids and cultivars for gardeners.

 It may seem curious that Monet, with his own ideas on art, should create a Japanese-influenced lily pond. However, he was greatly inspired by several Japanese artists whose paintings were impressionistic—but in a different style. "I had the good fortune to discover a batch of [woodblock] prints at a Dutch merchant's. It was in Amsterdam in a shop of Delft porcelain… suddenly I saw a dish filled with images below on a shelf. I stepped forward: Japanese woodblocks!" he wrote. The shopkeeper, unaware of their value, let him have them with the china jar. By the end of his life Monet had 231 in his collection. He liked prints of landscapes but rarely chose flowers and focused on the three major artists, Utamaro, Hokusai, and Hiroshige. This image and the pool and bridge at Ryōan-ji (page 7) make an interesting comparison.

In late May and early June, mauve and purple become prominent colors at Giverny. The subtle shift in shade between *Allium hollandicum*, Persian onion, and parrot tulips creates perfect pastel impressionism. Their differing habit has an impact, too: double-flowered tulips float like clouds, while the alliums have a stiffer, more compact form. Small contrasts such as this help to create texture in the garden. Ornamental alliums have untidy leaves so it is always useful to obscure them with ground cover plants—combining them with the lime green foliage and flowers of *Alchemilla mollis*, lady's mantle, is also effective.

Gertrude Jekyll

1843–1932

United Kingdom

Michaelmas daisy
Aster divaricatus

Jekyll planted this, with its clouds of tiny white daisies, among ground cover bergenias, adding a change of texture and color in late summer.

Gertrude Jekyll was born on November 29, 1843 and lived most of her life in the Surrey countryside, finally settling in 1897 at Munstead Wood near Godalming, with its 15 acres of woodland. A prolific writer and artist gardener, she created flower borders by "painting" with plants and established planting principles that are admired and copied to this day. Cottage-garden varieties were a particular favorite and she bred new strains of plants from seed. In a career spanning over 60 years she designed more than 400 gardens in Britain, Europe, and North America and wrote 14 books and over a thousand articles in journals and newspapers to instruct and inspire other gardeners. Her name is still uttered by gardeners in deferential tones.

Jekyll was a difficult child. Her father called her a "queer fish" and she infuriated her mother by clumping through the best rooms of their country house in her gardening boots. Although she was home-educated by governesses, a succession of eminent visitors from the world of arts and science (her mother studied music under Felix Mendelssohn, her father's interest in electricity and explosives led to a friendship with Michael Faraday) ensured a broad and stimulating education. She was described as "clever and witty in conversation, active and energetic in mind and body, and possessed of artistic talents of no common order."

In 1861, Jekyll enrolled at the National School of Art in South Kensington, where she was one of the first female students to study painting, receiving lectures on the principles underlying harmony in the composition of colors developed by Michel Eugène Chevreul. Her artistic style was influenced by the Impressionists. "Nobody has helped me more than Mr. [Hercules] Brabazon to understand and enjoy the beauty of color and of many aspects concerning the fine arts." She also admired William Turner, whose paintings she copied. Many compositional elements found in his paintings, especially the use of color, are reflected in her designs. She brought her ideas together in her book, *Color in the Flower Garden* (1908).

However, her talents were not confined to deft touches with an artist's brush; in time, she mastered a range of practical skills, among them carving, carpentry, gilding, walling, thatching, and gardening. She was an Arts and Crafts gardener well before the term was coined in 1888. Her gradual transformation from artist to artist-gardener was due to "extreme and always progressive myopia."

THE ARTIST AS GARDENER

Jekyll saw gardening as fine art, taking Impressionism from the gallery to the garden; plants became her palette. Breaking with Victorian formality, her flower borders incorporated daubs and drifts of color, the layout combining the structure of British landscapers such as William Nesfield with William Robinson's "wild gardening" style (see page 50).

She believed that "the first purpose of a garden is to be a place of quiet beauty such as will give delight to the eye and repose and refreshment to the mind." Although inspired by nature, she confessed that "no artificial planting can equal that of Nature, but one may learn from it the great lesson of the importance of moderation and reserve, of simplicity of intention, of directness of purpose, and the inestimable value of the quality called 'breadth' in painting. For planting ground is painting a landscape with living things, and as I hold that good gardening ranks within the bounds of the fine arts, so I hold that to plant well needs an artist of no mean capacity...for his living pictures must be right from all points, and in all lights."

Scottish thistle
Onopordum acanthium

This biennial needs well-drained fertile soil in full sun and makes a fine architectural plant. The Latin *onopordum* means "ass's fart" because of its effect on the digestive system.

There is no spot of ground, however arid, bare or ugly, that cannot be tamed into such a state as may give an impression of beauty and delight.

—Gertrude Jekyll

Jekyll was inspired by the natural vegetation in the countryside around her home and, somewhat surprisingly, by the plants, landscapes and architecture of the Mediterranean, having first visited Turkey, Rhodes, and Greece in 1863–64. She made several subsequent visits, using a specially designed pick to collect plants, which were sent back to England, where she trialled them for hardiness and usefulness as garden plants.

Most of her designs were collaborations. She created gardens for the War Graves Commission, private clients, local authorities, churches, schools and hospitals, and even a windowbox for an engineer in Rochdale; but her most successful collaboration was with architect Edwin Lutyens. He designed her home at Munstead Wood, completed in 1897. He provided the detailing for house and garden, she the planting plans that complemented the design and natural materials. The working relationship between Jekyll and Lutyens was summed up succinctly by Miss Jekyll: "…the difference between working with Nedi [Lutyens] and [Sir Robert] Lorimer was as between quicksilver and suet."

To ensure all her plantings were faithfully interpreted in the garden and no substitute plants were used, Jekyll started a plant nursery at Munstead Wood with the help of her Swiss gardener, Albert Zumbach.

Lamb's ears
Stachys byzantina

This plant is loved by children of all ages for its soft furry leaves. The cultivar 'Big Ears' is more robust than the species and remains tidier for longer.

Thousands of plants were dispatched annually: Sir George Sitwell's garden at Renishaw, Derbyshire, received 1,250 plants and more than 3,000 plants were sent to a client in Berkhamsted.

The gardens Jekyll designed have to a great extent been lost, although some have been restored: her own garden at Munstead Wood; Hestercombe, an outstanding example; and the delightful Old Manor House, Upton Grey, near Winchfield in Hampshire, a labor of love by Rosamund Wallinger. Jekyll's influence in style and use of color is reflected in the New Perennial movement of the early 21st century (see page 196).

Jekyll died on December 8, 1932. Three words on a gravestone in Busbridge churchyard, Godalming, Surrey, encapsulate her life and talents: Artist, Gardener, Craftswoman.

GERTRUDE JEKYLL

Gertrude Jekyll was a gardener and garden designer. She learned about plants by growing them, the only way to fully understand a plant and its habits. Below are her principles of garden design.

→ Create quiet spaces of lawn, without flower beds or any encumbrance.

→ Form simple groups of noble hardy vegetation, for beauty of flower, foliage, or general aspect.

→ Put the right plant in the right place; this involves technical knowledge and artistic ability.

→ Employ restraint and proportion in numbers and/or quantity by using enough yet not too much of any one thing at a time.

→ Create plant associations in sequences of good coloring, considering their form and stature and season of blooming, or of fall beauty of foliage.

→ Take care to join house to garden and garden to woodland.

→ Pulling plants forward: this was used to cover early plants after flowering, to correct mistakes when a nurseryman had delivered a short instead of a tall variety, and to vary the outline of a plant group. For example, when *Achillea filipendula* (yarrow) and *Eryngium oliverianum* (Oliver's sea holly) finished flowering, *Helianthus salicifolius*, the willow-leaved sunflower, would be bent forwards and tied over the space. It responded by producing a curtain of yellow flowers from the leaf axils along the stem.

→ Dropping in: plants of the correct color and size were grown in pots to fill gaps in the borders. A range was used, from hostas to lilies, but a favorite was pale pink hydrangeas.

Red hot poker
Kniphofia uvaria

This was one of the first red hot pokers to be grown in European gardens. It makes a wonderful splash of color in late summer and fall. Jekyll used these and others to soften shrub plantings.

The Lutyens-designed house at Munstead Wood was surrounded by Gertrude Jekyll's own garden, where she experimented with plants. Both the house and garden were built from local Bargate stone, in accordance with her Arts and Crafts principles, with simple sand paths between the borders. The planting consisted of seasonal and color-themed borders of hardy plants, including herbaceous perennials, shrubs, climbers, and hedges, with William Robinson's naturalistic informality at the woodland edge. Asters featured widely—there was a whole border devoted to them.

When a garden is divided, a doorway such as this one between two sections gains a magic of its own. The timber door and stonework, celebrating the creative skills of the Arts and Crafts movement, are transformed into a frame, with the garden beyond as the picture. Shafts of sunlight spotlight the silver foliage, creating a link between the two gardens and highlighting the ethereal, spiritual nature of the garden. Although the wall and door create a barrier between two differing areas and styles, the doorway is the temptation that entices the visitor onwards.

Henry E. Huntington

1850–1927

United States

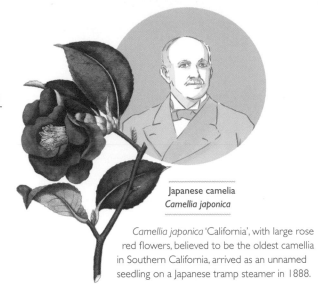

Japanese camelia
Camellia japonica

Camellia japonica 'California', with large rose red flowers, believed to be the oldest camellia in Southern California, arrived as an unnamed seedling on a Japanese tramp steamer in 1888.

Railway and real-estate magnate Henry E. Huntington was worth a fortune. His uncle's wife, Arabella Huntington, was equally wealthy; later in life, marriage brought their fortunes together. Both were collectors and they accumulated one of the greatest collections of art, rare books, and manuscripts in the world. Huntington was also an obsessive botanical collector. His 120 acre botanical gardens in San Marino, California, originally overseen by William Hertrich, with his expert organizational, landscaping, and horticultural skills, are now divided into more than a dozen specialized gardens.

The gardens and 15,000 varieties of plants are set within a parklike landscape of a country estate. Among the specialist gardens and collections are the Lily Ponds, Palm Garden and the world renowned Desert Garden. There is also a camellia grove, now recognized as an International Camellia Garden of Excellence, including nearly 80 different camellia species and some 1,200 varieties, many of them rare or historic cultivars.

In 1904, a year after buying San Marino Ranch, Henry Huntington was introduced to 26-year-old German William Hertrich, a landscape gardener who studied horticulture in Austria. Together, they searched local nurseries and visited other collectors to buy mature and unique specimens, collected seed, and imported rare and exotic plants from around the globe, turning Huntington's ranch into one of the finest botanical collections in the United States.

Hertrich started by building a nursery for mass plantings on the ranch. "I procured some seeds of redwood, and Incense and Himalayan cedars, of Canary Island date and other palms and propagated them…in a very few years we had…over 15,000 plants,"

Hertrich wrote. Next, he installed irrigation and experimented to discover which plants grew well in California, particularly commercial crops. Huntington sent seeds from melons he ate in France and Spain and returned home from a visit to his gentlemen's club in Los Angeles with avocado seeds: "He enjoyed them so much he asked the chef for all the seeds he had." This was the start of the first commercial avocado plantation in California, which grew alongside Huntington's commercial citrus, walnut, and cherry orchards, and more.

Now let us begin

The garden project began with the Lily Ponds; 4 acres were filled with two large and two small ponds, completed in 1904. They were fed by natural springs and heated pipes raised the water temperature; this extended the flowering of the lilies, particularly the giant Amazon waterlily, so the family could enjoy them in the relatively cool California winters. *Nelumbo nucifera*, the sacred lotus, was first planted here in 1905 and there is also *Cyperus papyrus* (papyrus) in the margins of the upper pond.

A palm collection was another early priority. Huntington admired their appearance and proportions and "he was particularly interested in the various types of cocos palms, because of their tropical appearance as part of the landscape."

Other palms were also trialled for their suitability as landscape material for home gardens, parks, and street and highway plantings. A specimen of *Phoenix canariensis*, the Canary Island date palm, was brought to the garden from the home of Huntington's Uncle Collis, which was destroyed in the 1906 earthquake; the tree is still scarred by the resulting fire.

The collection grew under Hertrich, despite several setbacks. In 1913, a cold winter with temperatures down to 41 degrees fahrenheit destroyed half the collection, most of them new or young specimens, which were particularly vulnerable. Disaster struck again during another severe winter in 1922; but as the trees matured they became

Date palm
Phoenix dactylifera

The date palm has been cultivated for its fruit since at least 4000 BC and is so widespread that its natural origin is unknown. Dates are eaten in sweet and savoury dishes, date nut bread, fermented to make alcohol, vinegar, and a host of other products.

HENRY E. HUNTINGTON

Huntington realized the need for a practical person to manage his gardening projects and was fortunate to find William Hertich; together they made a great team. Teamwork is essential for success in any garden; you will get by with a little help from your friends.

→ Henry Huntington collected seeds from avocados and started the first avocado orchard in California. It is worth saving seeds, particularly when travelling abroad; you may find something new on your plate or in the local market, even if they only become novelty plants. A man who went on an expedition to Mexico decided to germinate the seeds attached to his socks and introduced a new species to Britain, so anything is possible!

→ Huntington collected cacti and succulents after Hertrich mentioned he grew them as a boy. They are a wonderful introduction to plants for children. Most are slow growing and easy to care for. They need a bright sunny spot and regular but not excessive watering in the growing season. Reduce watering in winter and keep them in a bright, cool, frost-free place. When repotting, either wear thick gloves or fold a sheet of newspaper several times to make a thick strip, wrap it round the cactus and pinch the two ends together to pick it up.

→ Huntington wanted to grow the giant Amazon waterlily outside and had several ponds displaying lilies. If you are keen enough to heat the water, they can be grown outdoors in summer and kept indoors in winter in cooler climes. One or two gardens in Europe grow them that way, including *Hortus Botanicus* in Amsterdam.

→ Huntington's plants were sourced from nurseries, parks and gardens. If you visit a garden and like one of the plants, it is worth asking the owner for a cutting. Gardeners are generous; they may give you a rooted plant instead or choose the best cutting material for you. It may be tempting, but don't take cuttings without asking. If everyone did that, there would be no plants left in the garden!

Crown of thorns
Euphorbia millii

This extremely thorny plant from Madagascar is grown in gardens in warm climates and as a houseplant in cooler locations. It has been bred so that the brightly colored bracts now appear in different shades of red, pink, lemon—even mottled.

more cold tolerant. By the late 1930s, Huntington had sourced hardier palms in California, Europe, and Japan. There was an impressive collection of 450 palms (148 in the Palm Garden).

DESERT GARDEN

In 1907, Henry and Hertrich pondered the future of the barren eastern side of the property. Hertrich, who had grown cactus on his windowsill as a boy, suggested a desert garden as a solution; but Henry disliked cacti, remembering an experience from his days surveying the railway: "While backing away from some grading equipment that was passing by, he had his first painful and never to be forgotten introduction to the prickly cactus." (Hertrich, *The Huntington Botanical Gardens, 1905–1949.*) He was also unconvinced about planting on a large scale, but allowed Hertrich to plant a small trial garden on a baked, south-facing slope; it began with around 300 plants and gradually the collection grew. Hertrich bought cacti and succulents from parks and gardens and collected many from the deserts of California and Arizona. He landscaped the 10 acre garden with tons of volcanic rock. Aloes, agaves, and yuccas flourished and despite Huntington's early reservations, the garden's reputation spread among enthusiasts and his wealthy industrialist friends.

Today, the 12 acre garden—5,000 species of succulents and desert plants in 60 landscaped beds—is one of the oldest and largest in the world. It is massed with specimens planted where they grow best, where their differing appearances complement and contrast, such as the blue *Senecio serpens* (blue chalksticks), *Echinocactus grusonii* (the golden barrel cactus, with bright yellow spines), and the candelabralike *Echinopsis pasacana* (pasacana). These, together with large clumps of pincushion, barrel and other forms, create a pleasing, if somewhat surreal landscape. Many of the beds contain plants from the same country. *Cereus xanthocarpus*, weighing 15 tons, a 60 foot tall aloe, and a 100-year-old barrel cactus are among the stars.

Huntington and Hertrich's enthusiasm for plants and gardens never diminished. Hertrich stayed on as foreman and superintendent until 1948, continuing in an advisory capacity until his own death in 1966.

Huntington developed a plan to preserve his collections, intending that they should become a center for academic research that would grow and develop over time. Visit today and you will see that the collections and gardens are indeed flourishing, just as Henry Huntington intended.

The greater accomplishment was acquiring a great collection of British art, building a set of botanical gardens, and creating out of all of them a research institution for the benefit of human knowledge and culture.
—James Thorpe

◄ *Echinocactus grusonii*, the golden barrel cactus, and blue-gray *Agave parryi*, Parry's agave, are two iconic plants of the Desert Garden. Many of the golden barrel cacti in the garden were grown from seed before 1915 and now weigh several hundred pounds; others were collected from the wild. William Hertrich wrote, "I was selecting a few very desirable specimens of echinocactus for display purposes, selecting them for size, shape and color of their spines. When the Mexicans, who had been engaged to transport them from the desert by *burros*, found it difficult to handle the specimens because of their sharp spines, they put the machetes to use in cutting off the spines."

▲ This picture demonstrates the diversity in shape and form among cacti and succulents. Today the two dozen families of succulents and other arid zone plants have developed into a 10 acre garden display. It is the Huntington Garden's most important conservation collection, with more than 5,000 species of succulents and desert plants in 60 landscaped beds. All cacti are succulents but not all succulents are cacti; cacti are found almost exclusively in the Americas and are all in the family *Cactaceae*. Succulents fill their habitat in other parts of the world, such as Africa, and are found in a number of plant families.

Ellen Willmott

1858–1934

United Kingdom

Dutch crocus
Crocus vernus

This is the species from which most of the large Dutch varieties have been raised and selected. There are many named varieties in violet, purple, white, and stripes.

Ellen Willmott came from a family of keen and knowledgeable gardeners, who moved into Warley Place, near Brentwood, Essex, to develop their passion on a grand scale. On the deaths of her parents and godmother, Willmott inherited a vast fortune, which she spent developing three large gardens, two of which were in France. Lacking business acumen, she spent prodigiously on gardening, plants, and sponsoring plant hunters until her irrepressible passion finally swallowed all her inheritance. Today, her garden at Warley Place is derelict, with few remnants of its former grandeur, and is now a nature reserve.

Willmott's father was "something in the city," her mother and godmother independently wealthy, and the whole family, including her sister Rose, loved gardening. This passion prompted Frederick Willmott to buy Warley Place at Great Warley in Essex, where the oaks (*Quercus* spp.) and sweet chestnuts (*Castanea sativa*) were reputedly planted by diarist and arborist John Evelyn. They all set to work, laying the foundation for what would later be known as "the gardens which Miss Willmott's devoted skill made famous throughout the world,"

trumpeted an advertisement in *The Times* on April 18, 1935, and "not only one of the most beautiful, but also one of the most interesting of English gardens" (*The Garden*, William Robinson). Here, Willmott could be found hard at work in the meadow, orchard, kitchen and formal gardens, vineries, and glasshouses. Her mother, "a most energetic and enterprising gardener," was said to have raised a rose collection from seed and to have taken Ellen and Rose on an expedition, scouring the county for plants of Essex origin to plant in the formal garden.

Alpine Garden

In 1882, Willmott, age 24, started on her new alpine garden, using money she had saved from the £1,000 her godmother gave her each birthday. It was one of the first to be constructed on a grand scale and she employed James Backhouse of York, the most famous landscaping company of the day. Huge millstone grit rocks and boulders were transported from the North of England, and the excavations were dug deep enough to protect the plants from wind and rain. Pools, ravines with a stream, bridges, curved steps, and a glass-roofed cave and filmy-fern grotto emerged from the sculpted terrain, to be planted with treasures, many rare and difficult to grow, from New Zealand, the Andes, Greenland, Kashmir, California, the Cordilleras, and Tibet.

Willmott was hailed as a pioneer for her design, which was seen as a break from the "dreadful 'rockwork' of mid-Victorian times," William Robinson wrote. "…at Warley Place…we may see not only the rarest Alpine plants admirably grown, but effects and color not unworthy of the Alpine fields." Nearby was a small chalet, complete with herdsman's equipment and furniture, where Napoleon Bonaparte (with whom Willmott was obsessed) was said to have spent the night while crossing the Alps into Italy in May 1800, and which she bought and had rebuilt in her garden.

Plant passions

Willmott inherited Warley Place in 1898. Her finances were buoyant and her ideas extravagant, whether creating a gigantic boating lake on a hill, a hamlet of heated glasshouses, or an artificial gorge to display her prized collection of alpines. At the garden's peak, she employed more than 100 gardeners, who wore green-banded boaters, green silk ties, and navy blue aprons that she

Miss Willmott's ghost
Eryngium giganteum

The great gardener Ellen Willmott is said to have scattered seeds of this attractive spiny biennial, or short-lived perennial, species as her calling card in the gardens she visited.

herself had designed. They often found Willmott, who had "an innate love of flowers," in the garden with a basket and trowel, planting or weeding, when they arrived at 6am.

Willmott began hybridizing daffodils when she bought the stock of daffodil breeder Reverend George Herbert Engleheart of Appleshaw, Andover, amassing over 600 different species and hybrids. By the early 1900s she was winning prizes and Awards of Merit from the RHS for her new introductions, many named for her relatives and friends. Daffodils flowed in rivers and seas of yellow and cream over the vast lawns and under the trees of Warley Place.

After co-sponsoring the great plant collector Ernest Wilson's third expedition to China, Willmott turned her talents to seeds and bulbs, gaining a reputation for bringing his new introductions to RHS flower show standard more rapidly than anyone else. When Wilson visited in 1911, he was delighted to discover that she had been "wonderfully successful with seeds and plants which no one else managed to raise." Another plant collector whom she sponsored gathered rare pelargoniums from South Africa for her hothouses, while collectors from the nursery Van Tubergen's introduced bulbs from Armenia, Persia, and Turkestan, which were added to her garden.

Not content with Warley Place, where the plant collection is said to have numbered 100,000 different kinds, she bought two further properties: a chateau at Tresserve, near Aix-les-Bains, France, to continue indulging her love of alpines, and at Boccanegra, Italy, a haven for Mediterranean subjects. She only visited each garden twice a year for a month, at most.

Joining the RHS in 1894, she stood on several committees, showing regularly and winning a host of awards. She received the Victoria Medal of Honor and became one of the first three trustees of the new RHS garden at Wisley in Surrey.

Willmott lavished almost all her inheritance on her gardens. By the outbreak of war in 1914 she was bankrupt, eking out the rest of her life by mortgaging or selling her properties and belongings. After she died Warley was sold and the mansion demolished. The derelict garden is now managed by the Essex Wildlife Trust.

Willmott is remembered by many for her habit of scattering *Eryngium giganteum* seeds. It is known as "Miss Willmott's ghost" for its silvery appearance and because it would pop up in gardens she had visited. Gertrude Jekyll described her as "the greatest of living women-gardeners."

My plants and my gardens come before anything in life for me, and all my time is given up to working in one garden or another, and when it is too dark to see the plants themselves, I read or write about them.

—Ellen Willmott

ELLEN WILLMOTT

Ellen Willmott's garden at Warley was highly influenced by her friend William Robinson (she and Robinson attended the Chelsea Flower Show together in 1931) and his books, notably *The Wild Garden* and *Alpine Flowers for English Gardens*. The latter was into its second edition (1875) by the time Ellen and her family moved to Warley. "I have never seen anything more beautiful in nature or gardens than grassy banks planted with the smaller and rarer Narcissi in the gardens at Warley Place," Robinson wrote.

→ "The gardeners' children were persuaded to throw handfuls of bulbs from a wheelbarrow over the ground and they were planted where they fell," notes Audrey le Lièvre in *Miss Willmott of Warley Place: Her Life and Her Gardens*. One order is said to have been 10,000 bulbs. This creates a "naturalistic" effect; crocus, camassias, and daffodils and species such as *Narcissus bulbocodium* (the hoop petticoat daffodil) can all be planted that way.

→ As a plantsperson and botanist, Ellen Willmott collected specific groups of plants, publishing a book called *The Genus Rosa* using examples from her own garden. This is not about using plants for their collective artistic merit but growing them for their individual beauty and history or rarity, which creates another "layer" of interest in the garden.

→ A well-constructed alpine or rock garden should reflect the natural strata seen in the wild—it should not be a pile of soil with rocks sticking out. Visit mountainous locations and take photographs for inspiration, noting how rocks lie next to one another. They may be almost buried with only a small percentage protruding from the soil and small pebbles, or shale, forming rivulets between the pieces. Recreate this using several large pieces and the impact will be natural and impressive.

Hardy plumbago
Ceratostigma
willmottianum

The great plant-hunter Ernest Wilson found this in China in 1908. Miss Willmott raised two plants from the seed and from these specimens came most of the early plants in Britain.

Edward Augustus Bowles

1865–1954

United Kingdom

Early bulbous iris
Iris reticulata

This well-known early flowering iris, appearing at the end of winter, is available in many color forms. It needs an open, sunny, well-drained site.

Edward Augustus Bowles ("Gussie" to his friends) was an artist, entomologist, plantsman, philanthropist, writer, and arguably the greatest amateur gardener of his era. He developed a garden at Myddelton House, Enfield, Middlesex, and was passionate about gardening and collecting, breeding and growing plants, often from specimens that he found in the wild. He had a penchant for crocus and snowdrops (coining the term "galanthophile" for snowdrop enthusiasts), but he also had an eye for the quirky and rare. A dedicated churchman, he devoted much of his time to charitable works and was kind, generous spirited, and humorous; both garden and gardener remain influential.

After almost losing the sight in his right eye at the age of eight, Bowles was educated at home. He later went to Jesus College, Cambridge, to read theology, intending to become a clergyman, but on the deaths of his elder brother and sister within three months of each other, he returned home to comfort his parents. There he pursued his interest in natural history, gardening and painting, and devoted himself to the Church. A friendship with the keen gardener and scholar Canon Henry Nicholson Ellacombe (1822–1916), "the foremost both in time and ability of my teachers in garden-craft," developed Bowles's interest in horticulture.

In 1893, he began to construct a rock garden; the gravel subsoil and low rainfall were ideal for crocuses and other bulbous plants, a particular area of interest to him. Having been inspired by past trips to France and Italy, in 1898 he travelled to Malta, Egypt, Italy, and Greece. A hay fever sufferer, Bowles made regular summer visits to the

Alps to gain respite, sometimes with friends, including plant collector Reginald Farrer, the "father of twentieth-century rock gardening." Here, as at home, he painted beautiful, detailed watercolors despite his visual handicap, observing plants and sending back species that flourished in dry soil.

GUSSIE'S GARDEN

Bowles' travels inspired the creation of an alpine meadow at Myddelton House. A mass of snowdrops, crocuses, daffodils, and camassias still appear there in spring, followed in summer by a cloud of blue geraniums. In summertime, Bowles was often to be seen dressed in his blue and white striped Edwardian bathing costume and straw boater, wading into the pond to clear it of blanket weed.

Bowles was religious and compassionate, and showed great kindness to the poor of the parish. He ran a night school to teach local boys to read and write; they came to work in the garden and were known as "Bowles boys." At weekends they were allowed to take part in organized activities such as football, cricket, and fishing, and skating in winter. It was they who collected the stones for the Stone Garden, which was based around the trunk of a fossilized tree found during the excavation of the King George V Reservoir at Chingford.

RESPECTED WRITER

In 1912, the editor of *The Gardeners' Chronicle* encouraged Bowles to write about his garden through the seasons. This became his celebrated trilogy: *My Garden in Spring* (1914), *My Garden in Summer* (1914), and *My Garden in Autumn and Winter* (1915), bestsellers in their day and still available now. Reginald Farrer wrote that they showed "how a gentleman can wear his garb of knowledge with a gay air and humor, dignified yet easy, and whimsical and personal."

A Handbook of Crocus and Colchicum for Gardeners followed in 1924, based on his own experiences of growing, observing and painting crocuses, his "first garden love" (he was dubbed "the Crocus King"). It was the major reference work for many years and was followed by *A Handbook of Narcissus*

Snowdrop
Galanthus nivalis

This species has been widely cultivated in gardens for centuries and is very variable; there are double forms and a range of different markings, including some with yellow patches on the flower segments instead of green.

EDWARD AUGUSTUS BOWLES

Edward Augustus Bowles was generous, so many of his lessons are about giving and sharing, a familiar gardeners' trait. He gave so much of his time, knowledge, and enthusiasm, and so many of his plants to others, it is little wonder that he was so widely loved and respected.

→ Share plants with friends so that if yours dies, you have a source of replacement. "But I have never yet sold a plant, and I hope I never shall…" (*My Garden in Summer*). This is particularly important with anything rare and unusual. Don't give away plants that self-seed freely, as they can become a menace and you'll be remembered for the wrong reasons.

→ "I only wish to show how much pleasure I have derived from this by no means remarkable garden, and to encourage others somewhat similarly circumstanced to collect and grow plants of all kinds, to watch and note their peculiarities, mark their charms, and hand on the best of them to others who love a plant for its own sake" (*My Garden in Autumn and Winter*). Bowles gained pleasure from a wide variety of plants. Try growing something new in the garden, adding a number to your collection every year. This may lead to a new interest or to a group of plants that flourish in your garden.

→ "Do all that you know, and try all that you don't…in choosing a place for a plant…" (*My Garden in Summer*). If a plant doesn't thrive, try moving it carefully to another part of the garden, several times if necessary, until you find a place where it thrives. Knowing its natural habitat makes the decision easier.

→ Gertrude Jekyll visited Myddelton House to meet Bowles and look round the garden. A guided tour of any garden is an ideal opportunity to learn about its plants and how to grow them.

Ivy leaved cyclamen
Cyclamen hederifolium

This is such an easy plant to grow. It is very hardy, thrives in sun or semi-shade, and seeds itself freely when happy. The flowers in pink or white are followed by attractively marked leaves continuing over winter. Individual corms can be very long lived.

(1934). His work on *Anemone*, with W. T. Stearn (1911–2001), was incomplete at the time of his death in 1954, having been abandoned due to lack of funding, but a final book on snowdrops, in collaboration with Sir Fredrick Stern (see page 106), was published posthumously in 1956 under the title *Garden Varieties of Galanthus*.

In May 1897, Bowles became a life member of the RHS and would serve diligently and skilfully on 15 committees. He was on the council for 36 years, serving as vice president from 1926 to 1954. The society awarded him its highest honor, the Victoria Medal of Honor, in 1916, and a Gold Veitch Memorial Medal in 1923 for his work on crocuses and other bulbs. They also asked him to design the Grenfell Medal, first awarded in 1898, to be awarded for an exhibit of paintings, drawings, or photographs, which he then won on five occasions. The RHS garden at Wisley still has a Bowles' Corner, using plants from Myddelton House (though much of the original planting has been lost), which was renovated during 2014, the 60th anniversary of his death.

SOMETHING COMPLETELY DIFFERENT

Bowles was an avid collector and selector; he was always on the lookout for variegation and mutation, embracing the rare, unusual, and downright weird. All were grown in two areas of his garden, known as "Tom Tiddler's Ground" and the "Lunatic Asylum," containing the likes of *Fragaria*

To sum up the present conditions of the garden; climate, soil and trees contrive to make it the driest and hungriest in Great Britain, and therefore arises the line of gardening I have been driven into. It is perhaps better described as collecting plants and endeavouring to keep them alive, than as gardening for beautiful effects or the production of prize-winning blossoms.

—Edward Augustus Bowles,
My Garden in Spring

vesca 'Muricata', the "Plymouth strawberry" (discovered in 1627 by John Tradescant), and *Plantago major* 'Rosularis'.

More than 40 plants raised by Bowles are still available; others were named for him. *Carex elata* 'Aurea' (Bowles' golden sedge) he found at Wicken Fen, Cambridgeshire; he discovered *Vinca minor* 'La Grave' (syn. 'Bowles Variety' and 'Bowles Blue') in the churchyard at La Grave, France; and *Milium effusum* 'Aureum' (Bowles' golden grass) came from Birmingham Botanical Gardens. Though he raised many crocus hybrids, very few remain.

Bowles died at Myddelton House, in his favorite room overlooking the garden, on May 7, 1954, a week before his 89th birthday. Three hundred people attended his funeral and his ashes were scattered on the rock garden, his favorite part of the garden. Many remember him today as "the kindest of old souls" and "a great gardener."

Pierre S. du Pont

1870–1954

United States

Flame azalea
Rhododendron calendula

So-called because the buds resemble candle flames. It is found in the Appalachian Mountains.

Pierre S. du Pont was born in 1870 in a house overlooking Brandywine Creek near Delaware. He was an industrialist by profession and a gardener by vocation. After buying a small farm about 10 miles north near Kennett Square, Pennsylvania, he developed a theatrical flair for horticulture, inspired by gardens he had seen on his travels. It is said that most of the garden styles of the Western world, from the Renaissance onwards, are present at Longwood, and it is the ultimate expression of an American country estate garden of the 1920s. He intended his garden to be shared and it is still a center of excellence today.

In 1798, twins Joshua and Samuel Peirce started one of the finest arboreta in the USA. After the death of heir George Washington Peirce in 1880, it gradually fell into decline; Pierre S. du Pont bought the land to prevent the trees from being felled for timber. "I have recently experienced what I would formerly have diagnosed as an attack of insanity; that is, I have purchased a small farm," he wrote to a friend in 1906, adding, "I expect to have a good deal of enjoyment in restoring its former condition and making it a place where I can entertain my friends."

Du Pont's extensive travels influenced his ideas on garden design. He was impressed by Italian and French formal gardens with their geometric structure and form, visited the Horticultural Hall in the Moorish style at the 1876 Centennial in Philadelphia and appreciated Joseph Paxton's Crystal Palace at Sydenham and the glasshouses at the Royal Botanic Gardens, Kew. He also took in as much natural beauty and as many gardens as possible during trips to South America, the Caribbean, Florida, California, and Hawaii. His garden became a fusion of inspirations.

There was no masterplan. "Whatever Mr. du Pont did in construction or addition came out of his mind at the time," recalled a staff member at Longwood. "When he built the greenhouses in 1920, he never thought for one minute that nine years later he would put waterfalls and fountains in the corn fields in front of them."

Du Pont began creating the garden in 1907, designing and drafting all the outdoor gardens, including the fountain gardens, until the 1930s. He disliked imprecision; having commissioned landscape architects for a design round a former home, he found errors in their survey, dismissed them immediately, and vowed to do the work himself. His first project was the 600 foot Flower Garden Walk, with the first fountain—a pool with a single jet of water—in the center of the walk, his favorite plants, cottage-garden flowers, trellises burgeoning with roses, picturesque benches and a birdbath. He noted: "I have set myself and guests to work planting flower seeds whenever I have opportunity." In 1910, he hired gardeners because he was so busy running the DuPont Company (the garden and working farm were expanding rapidly), though he often did the designs.

In 1909, he began hosting garden parties that rapidly became a highlight of the "summer season." Their success encouraged him to search for more wonderful ways to enthrall his guests. In 1913, he and his future wife visited 22 Italian villas in search of Renaissance inspiration and found it in an outdoor theater at Villa Gori near Siena. It was first used in 1914; within a year, he added "secret" fountains that shot out of the stage floor to drench visiting nieces and nephews, having seen *giochi d'acqua* in Italy. In 1926–27 he redesigned the theater with underground dressing rooms and 750 fountain jets, including a 10 foot high water curtain illuminated by 600 colored lights.

He also connected the new and old wings of the original house with a conservatory, the first "winter garden." Its courtyard displayed plants with exotic foliage and a small marble fountain, a wedding gift marking his marriage to Alice Belin in 1915.

May flowered orchid
Laelia speciosa

This orchid with wonderfully vivid flowers is found in arid zones of Mexico, in cool to cold conditions. It is drought resistant and needs plenty of light and a cool winter rest.

By 1916 he contemplated a grander winter garden, "designed to exploit the sentiments and ideas associated with plants and flowers in a large way," he wrote in a letter to architect Alexander J. Harper. Longwood's conservatory, in the Palladian style, opened in 1921, covered 3.5 acres, and was built to shelter garden party guests from the rain. It was also to be a perpetual Eden for ornamental horticulture and production of fruits and vegetables. Although the head gardeners designed most of the planting, Mr. du Pont no doubt approved the plans and was certainly involved with their selection, visiting nurseries in California to order plants. He decided to fill his conservatory with decorative displays rather than follow the fashion of using exotic tropical plants.

A newspaper article published in October 1926 noted: "The Orangery had green grassy planting beds filled with orange and grapefruit trees. Formosan conifers towered above the citrus. Bordering the citrus trees were subtropical plants, including: cypress trees; mimosas; tree ferns; banana; coconut palm; coffee; guava; kumquats; mango; papaya; begonias; fuchsia; heliotrope; lantana; orchids; *Plumbago propensus*; and South African violets. Azaleas bloomed in spring, chrysanthemums in the [fall], and poinsettias at Christmas."

The fundamental aim of the place is to do everything that is attempted in a first class way.

—Pierre S. du Pont, 1912

FOUNTAINS FOR LIFE

Inspired by Villa Gamberaia near Florence in Italy, du Pont built a Water Garden, incorporating 600 fountains in nine varying displays, spouting from six blue-tiled pools and 12 pedestal basins. Perfectly enclosed by clipped lindens on the sides and evergreens at the far end, it is reminiscent of enclosures in the woods at Versailles. At the same time, he installed a jet fountain with a 40 foot spout at the end of the *central allée* in Peirce's Park. But his masterpiece was the 6 acre Main Fountain Garden, with canals, fountains, and statuary, inspired by the hydraulic spectacle he had seen at the 1893 World's Columbian Exposition in Chicago as well as by the fountain gardens of Italy and France. More than 380 fountains and spouts in canals and pools using 1,000 gallons of water a minute blast water up to 130 feet into the air. Du Pont did many of the hydraulic calculations himself. At night they were illuminated by red, blue, green, yellow, and white lights to form infinite color possibilities.

Pierre S. du Pont continued buying land and shaping his masterpiece until his death in 1954, spending about $25,000,000 on it over his lifetime. He was awarded the George Robert White Medal of Honor in 1926, from the Massachusetts Horticultural Society, the highest horticultural award in America, and the Gold Medal of the Horticultural Society of New York in 1940.

Today, Longwood is one of the world's premier display gardens, renowned for its horticultural excellence and attention to detail. It is still a garden for all to share.

PIERRE S. DU PONT

Longwood, one of the world's premier display gardens, is still renowned for its excellence in horticulture, attention to detail and, as du Pont wanted, it is still a garden to share. It is also the ideal garden to view many classical garden styles in one location.

→ Du Pont was enthusiastic about water features. Ponds, fountains, ripple pools, and waterfalls add another dimension to the garden; they can be exciting, soothing, or cooling, according to your wishes. You could try something similar to *giochi d'acqua* at home. It could be permanent, temporary, or perhaps to make a party memorable and fun.

→ Longwood boasts a magnificent theater and is a place to entertain. Why not think laterally and use your garden for other pastimes—outdoor cinema against a white wall, croquet, chess, garden games, even a boule court adds another dimension to your recreational space.

→ Du Pont wrote of wanting his garden to be "a place where I can entertain my friends." When designing your garden, always incorporate dining space. It should be long enough for a table and wide enough so you can comfortably pull out the

chairs without falling into the flower bed or onto the lawn; take your measurements accordingly and make sure they are correct.

→ A conservatory, however small, can be turned into a winter garden. Add winter-flowering plants, even if they are only temporary displays in pots, to brighten the winter and hasten the spring. You can enjoy their cheering color from the shelter of the conservatory, making you feel as though you are in the garden, even on the coldest day of winter.

→ Du Pont worked his gardens himself, but as time pressures and the size of the garden increased, he employed staff to help. Your garden should be a pleasure, not a pressure; get help if you need it, even to trim hedges or mow, allowing you to make time for more interesting jobs.

St Joseph's lily
Hippeastrum x johnstonii

These vibrant red flowers have a spicy fragrance. It was believed to be the first *Hippeastrum* hybrid ever produced, from a cross made by Arthur Johnson, a watchmaker from Prescot, Lancashire.

There are 20 gardens in
the 4 acre conservatory
at Longwood Gardens,
including the Mediterranean,
Orchid, and Palm houses. The
Exhibition Hall and Orangery, the
centerpiece of the conservatory
that was built in 1921, hark back
to the elegance of a bygone era.
Bougainvillea from the original
planting is trained on the walls
and pillars and the flooded marble
floor acts as a mirror, reflecting the
floral displays. In Pierre S. du Pont's
day it was used for elegant dinner
parties and dances; it is still drained
for events and performances today.
Among the notable plants on
display are *Howea forsteriana*, the
kentia palm, *Cyathea cooperi*, the
scaly tree fern, and *Bougainvillea
glabra* 'Penang', with bracts of
pink-purple.

Lawrence Johnston

1871–1958

United Kingdom

Flame creeper
Tropaeolum speciosum

This is often planted so it scrambles over evergreen plants or up yew hedging, as at Hidcote. It was introduced through Veitch's nursery, by William Lobb.

Lawrence Johnston, an American born in Paris, came from a well-connected, wealthy family. A plant collector and artistic man of vision, he created two great gardens in contrasting locations and styles: Hidcote, in the windswept Cotswolds of England, and Serre de la Madone, on a sun-baked slope in the South of France. He filled both with theatrical plantings and choice rarities, many collected while plant-hunting abroad. Of all his achievements, his greatest legacy is Hidcote Manor Garden, said by many to be one of the finest gardens in the world.

In 1907, when Johnston was 36, his mother, Gertrude, bought the small estate of Hidcote Bartrim in the Cotswolds. With great vision, energy and zeal, he went on to transform what was open windswept farm-land, with a clump of beech trees and a cedar of Lebanon, into one of the most influential gardens in the world. Johnston was meticulous in his preparation. He studied architecture to restore the Hidcote Manor house and buildings, then turned to garden design to restore and enlarge the garden.

He began laying out the "bones" of the garden in strong formal lines, using hedging to create wide vistas, small "rooms" and spaces, and to provide structure and shelter. It appears there were no plans; he pegged out ideas on the ground and the garden evolved. Vita Sackville-West (see page 122) described his design as "a series of cottage gardens" and complimented his use of texture in the mixed hedging, likening one with five different things in it—yew, box, holly, beech, and hornbeam—to "a green and black tartan."

Each of the spaces was themed and unique in design, scale, color, and atmosphere, hence their names: the

Bathing Pool Garden, the Fuschia Garden, the Pillar Garden, the Red Border Garden, each one a sharp contrast and surprise. There is an intimate White Garden, framed with dark hedging and topiary, overlooked by a small thatched cottage, and a Stream Garden, a relaxed area of spinneys and ditches filled with lush leafy moisture-lovers such as hostas and *Lysichiton americanus*, the skunk cabbage, with its bright yellow malodorous flowers and huge paddle-shaped leaves. From here, you step into another extraordinary space—the majestic vista of the Long Walk and its two parallel hedges. The whole garden is filled with theatrical and architectural statements and endless

surprises, using texture, form, and color; a mix of plantsman's rarities, such as *Berberis temolaica* (a barberry), with purple stems and blue-gray leaves, and refined, artistic planting of the highest order.

SERRE DE LA MADONE

Not content with making one great garden, Johnston created a second, at Serre de la Madone near Menton on the French Riviera. He bought a farmhouse in 1924, purchasing land over several years until he owned 25 acres. It was here he spent his winters from September to April, finally settling permanently in 1948 when he gave Hidcote to the National Trust, which secured its future.

Windswept spaces dictated the design at Hidcote; at Serre de la Madone it was the topography. He divided the steep, terraced, southwest-facing slope with a central stone path. Connected by several flights of steps from this axis was a series of terraces of different heights and dimensions, supported by dry-stone walls. These followed the curves of the terrain from the more formal part of the garden, the planting softening as it merged with the countryside beyond. In one area he used berberis, following the undulations of the site, which was lifted, according to the garden designer Russell

Yucca-leaved beschorneria
Beschorneria yuccoides

This spectacular Mexican plant, with its unusual stems and flowers, is perfect for creating an exotic effect and is a "must have" plant in Mediterranean gardens.

Page (1906–85), "onto a higher plane [by] tufts and groups of Yuccas, *Y. flaccida, Y. filamentosa* and *Y. gloriosa.*" Within these terraces were several features: a Moorish Garden, Rockery, Plane Garden, and—five levels below the house, on the largest terrace—a serene rectangular pool and conservatory.

A PLANTSMAN'S PARADISE

The benign climate was ideal for growing Mediterranean and subtropical plants and Johnston the plantsman was eager to capitalize. Here he grew tender beauties such as *Rosa chinensis* 'Bengal Crimson', the Bengal rose, with single cherry-red flowers; exotic Australian banksias, bottlebrushes, and tea trees; and cycads and succulents from all over the world.

He also satisfied his instinct for theater with bold, spectacular plantings: a box-edged parterre, for example, carpeted with purple periwinkle with *Tulipa clusiana*, the red-and-white-striped "lady tulip," growing through it. Elsewhere, on a slope, pink trumpet-flowered *Amaryllis belladonna* (Jersey lily) grew through a carpet of azure blue *Ceratostigma plumbaginoides* (a hardy plumbago); one terrace was planted with nothing but *Beschoneria yuccoides* (yucca-leaved beschorneria), with bright pink stems and flowers.

Johnston also travelled to Java, South Africa (with Major Collingwood "Cherry" Ingram) and Yunnan in search of plants, the latter with the great plant collector George Forrest, a trip that was curtailed mainly owing to Johnston's ill health. However, it did yield pale lavender *Iris wattii*, which Johnson dug up near an irrigation channel in Yunnan, and deep-yellow-flowered *Mahonia siamensis*, collected by his bearer (he was carried in a sedan chair). All of these and more were planted in his gardens and shared with his plant-enthusiast friends.

Among his special plants were *Stenocarpus sinuatus*, the "Firewheel tree," with bright red flowers, from the rainforests of Australia's warm east coast, and *Otatea acuminata* ssp. *aztecorum*, the Mexican weeping bamboo, with its fine-leaved, feathery form. If this was not enough, he built a vast 3 acre aviary, where a collection of exotic birds flourished in semi-captivity.

Johnston was honored by the RHS in 1947 with a Gold Veitch Memorial Medal, in appreciation of his vision and creative genius, from a grateful gardening world.

There has been no more beautiful formal garden laid out since the time of the Old Palace at Versailles than that designed on a small sale, but with exquisite artistry by Major Lawrence Johnston [at Hidcote]. Not only that but the garden is filled, as earlier gardens were not, with interesting and beautiful plants, some of which he himself collected in the mountains of China.

—2nd Lord Aberconway

LAWRENCE JOHNSTON

Lawrence Johnston's two gardens had some similarities: rare and unusual plants sharing space with theatrical ornamentals, bold plantings such as the Red Borders, and the massed plantings of *Amaryllis belladonna*, the Jersey lily. Every garden should have something with a "wow" factor, be it rare plants or bold plantings or even unusual statuary.

→ Garden rooms create shelter and spaces and bring an element of surprise. This can be achieved using trellising, fence panels, or hedging.

→ One of the features at Hidcote is the formal hedging with its sharply clipped lines, creating a strong structure. This is easier to achieve using yew hedging (*Taxus baccata*) because of its small leaves and dense foliage. It casts a dense shadow and should be positioned with this in mind. It also creates a plain, dark-green background, highlighting colors in the borders. When planting hedging, do remember that it has to be trimmed, so don't plant more than you can care for.

→ Most gardeners use only one variety when planting hedges. Lawrence Johnston used a range of varieties to create color and texture. Vita Sackville-West appreciated this, noting how "the flatness of the yew, contrasted with the interplanted shine of holly." The hedge itself becomes an ornamental feature.

→ Terracing was used to create planting space on the steep slope in his garden at Serre de la Madone, the width of the terraces turning most of them into small gardens. Where space is at a premium and beds are more like a series of steps, put a narrow access path at the back of each terrace, so you can reach into the planting area above to make gardening easier.

→ In both his gardens, Lawrence Johnston designed formally near the house, gradually becoming more naturalistic as the gardens reached out towards the countryside; the garden had an empathy with the surroundings. Plantings in smaller gardens can be designed to link with what grows beyond, making the garden feel larger and borrowing the landscape.

→ Johnston established several principles for himself. "Plant only the best forms of any plant" was primary. "Plant thickly" was another, in the knowledge that where the gardener doesn't put a plant, nature will.

Firewheel tree
Stenocarpus sinuatus

This Australian rainforest tree and member of the *Protea* family has proved to be adaptable in cultivation.

Dividing a garden into rooms adds an element of surprise, prompting visitors to wonder what lies beyond. It is also a way of making a garden seem larger, as you take time to meander slowly, absorbing the details of each room. The Maple Garden, part of the old garden at Hidcote, with its collection of Japanese maples and plantings of *Heliotropium arborescens* 'Lord Roberts', silver-leafed *Centaurea gymnocarpa*, the velvet centurea, and the pale cream-flowered *Hydrangea arborescens* ssp. *discolor* 'Sterilis', smooth hydrangea (a rare form of a tough, North American shrub), is a perfect example of how rooms transform a large garden into something more intimate.

Beatrix Farrand

1872–1959

United States

Winter heath
Erica carnea

A widely planted reliable dwarf shrub, winter heath forms dense mounds or mats and is good for suppressing weeds.

Landscape gardener Beatrix Farrand was born on June 19, 1872 into a family boasting five generations of plant lovers. Later in life, a chance meeting with the wife of Professor Charles Sprague Sargent, director of the Arnold Arboretum of Harvard University, led her into landscape design. A tour of European gardens ignited her enthusiasm and a series of grand commissions, including the White House, Dumbarton Oaks, and University campuses at Princeton and Yale, sealed her fame. She married Max Farrand, director of the Henry E. Huntington Library and Art Gallery, California, and created gardens of her own, notably at Reef Point and Garland Farm in Maine.

Farrand enjoyed gardening with her grandmother, who owned one of the first espaliered gardens in Newport. Aged eight, she observed the laying out of her parents' summer retreat at Bar Harbor, Maine; she learned the names of roses and was taught to "deadhead" them. As her age increased, so did her enthusiasm.

Farrand's mother was friendly with the wife of Professor Charles Sprague Sargent, director of Arnold Arboretum. Noting Farrand's love of plants, Sargent suggested that she study landscape gardening. She went to live on their estate, Holm Lea in Brookline, Massachusetts, where Sargent taught her the basic principles of landscape design, including turning ground plans into reality "to make the plan fit the ground and not to twist the ground to fit a plan." She was a conscientious, observant, and critical student, remarking of Boston Gardens, "The planting is very bad. There are many fine trees in the Garden but they do not show up for much as they are not led up to in any way, nor is the multitude of small beds calculated to give repose to the eye and give breadth of effect."

Gardening influences

Sargent advised Farrand to travel widely, to look at gardens and learn from what she saw, especially in Europe; she often did so with her aunt, the novelist Edith Wharton, visiting France, Italy, Germany, the Netherlands, England, and Scotland. Her observations are recorded in her notebook— her *Book of Gardening*. She met Gertrude Jekyll at Munstead Wood (see page 62), where she learned about painting with perennials (she later acquired her archives of over 300 plans, photograph albums, and plant lists), and visited the gardens at Penshurst Place in Kent; both experiences left a lasting impression on her. European gardens became the reference point for her designs, which sat between "the formalists on the one hand and the naturalists on the other," emphasizing the relationship between the garden and the house and seeing gardening as art. "She was outstanding as a designer, she had a great sense of proportion and strength in her designs…because she was a woman, she didn't have the opportunities that men had; while they designed public parks, hers were in the private realm." (Diana K. Maguire, landscape architect.)

White spruce
Picea glauca

This slow-growing species often lives for many years. The northernmost tree species in North America, it is extremely hardy and can survive in exposed locations.

Reef Point

All the experience and understanding that she gained from her travels is encapsulated in her own gardens. She designed them in a formal structure but planted in the "natural" style of William Robinson (see page 50), whom she visited while in England. In her plantings at Reef Point, she also added asymmetrical drifts of different genera in their scientific classifications, a trial garden of native plants, and a collection of single roses, the largest in the country. Her interest in plants and gardening was a combination of art and science.

In order to create an illusion of space, which she had seen in 18th-century English landscapes, she intersected the vistas with curved paths and subdivided the garden using different plantings: laurels and rhododendrons on the way to the vegetable garden, an orchard of dwarf apples, and groups of red and white spruce planted as windbreaks, all of which unfolded to visitors as they followed a prescribed route. There were vistas over purple heathers to blue water between pointed firs, borders of heliotrope by the porch, and white nicotiana along the path, providing radiance by day and fragrance at night. Farrand's own hands-on gardening was "tussling and planting around the rocks." Her passion for this ran deep, because of her

Southern magnolia
Magnoilia grandiflora

Thrives against a sunny wall in cooler climates or as a freestanding tree or large shrub in warm climates. Its flowers can grow up to 10 inches across.

childhood fascination with the plants on Mount Desert Island, where Reef Point is situated, and she described herself as "a barbarian for digging up wild plants" (a practice that is now illegal).

A FIRM HAND ON AN UNCERTAIN FUTURE

In the winter of 1954, Farrand, now a widow aged 82, reviewed her circumstances; costs were increasing and there was little chance of tax exemption, an essential factor if her dream of creating a Foundation and School of Horticulture and Design at Reef Point was to become a reality. With this and her advancing age in mind, in 1955 she decided that if her garden at Reef Point could not be maintained to her standards, she would rather it were destroyed. Some of her associates were incredulous but,

A taste of gardening can add so much to life…intimate contact with growing things…gives interest and flavor to each day.

—Beatrix Farrand

"Mrs. Farrand was as determined now to put an end to the garden as she was to create it." Both house and garden were dismantled and two new civic gardens were created from her plants.

Farrand gave her library of 2,700 books, 1,800 herbarium specimens, and hundreds of garden plans, including the Gertrude Jekyll archive, to the University of California, Berkeley, and only took the things she loved to her new home, Garland Farm on Mount Desert Island in Maine. There were artifacts from Reef Point, including a door, her treasured collection of seed packets from plant societies and botanic gardens around the world, and favorite plants including the Sargent cherry, *Metasequoia*, heathers, heaths, and herbaceous perennials—she even exhumed the remains of her pets. Garland Farm was designed so she could view the terrace from her room and see her favored blues, purples, greys and a sea of heather, with pruned cherries and shrub roses running down toward a meadow. It was a last small creation by a great landscape gardener.

To demolish the house and garden at Reef Point seems drastic. Yet despite its memories and her hopes for the future, it was just one of many that Farrand created. Most gardeners put their creative energy into one or two gardens and become emotionally attached, but Farrand realized the ephemeral nature of her work. "Written words and illustrations outlive many plantations," she once wrote.

BEATRIX FARRAND

Beatrix Farrand wanted to be known as a landscape gardener. She saw a practical link between designing, building, and maintaining. Maintenance is an essential part of any gardening project; it is the skill of a practical gardener that maintains the original design.

→ Keep climbers well pruned and trained, so they don't swamp the support they are growing on. This requires that the garden be constantly observed and managed. Never be without your pruners! Using this technique, Farrand ensured that the line of the trellis or wall that plants were growing on still contributed to the structure of the garden.

→ Be aware of the relationship between the cultivated and natural aspects of your garden. The scalloped edges of Farrand's flower borders at Reef Point mimicked the edge of the forest beyond the garden, so the cultivated blended seamlessly with the natural.

→ When designing a garden, work with the existing topography, rather than imposing the garden on the landscape.

→ Farrand maintained exacting standards of maintenance, ensuring that the detailing of her planting—the "hard" landscaping, such as paths and features, seats, and statuary—were all displayed to perfection.

Creeping dogwood
Cornus canadensis

Unlike most dogwoods, which are large trees or shrubs, this pretty herbaceous perennial groundcover is ideal for moist acidic sandy peat or leaf mould. The white petals are actually bracts.

▶ This beautifully designed and maintained path at Dumbarton Oaks, leading to a statue of Pan and the Lovers' Lane Pool, winds sinuously through a mixed planting of *Epimedium*, bishop's hat; *Narcissus*, daffodil; *Malus*, crabapple; and *Buxus sempervirens* 'Suffruticosa', box. Note the pattern in the brickwork; the quality of landscaping materials is as important as the planting when creating a garden. Beatrix Farrand's gardens were all finished to an exceptionally high standard.

▲ Wisteria is used extensively by Beatrix Farrand at Dumbarton Oaks, where it cloaks pergolas and trellis screening. This *Wisteria floribunda*, Japanese wisteria, above a low stone wall at the north end of the North Vista, encapsulates the simple elegance of her designs. Wisteria is an extremely versatile plant: it can be grown as a small tree, trained as an espalier, or even as a single stem along a balcony, where "less is more," or allowed to scramble into a large tree as it would in the wild. It always looks beautiful against a background of brick or natural stone.

Henry Duncan McLaren

1879–1953

United Kingdom

Chilean firebush
Embothrium coccineum

This species is among the most desirable of all garden plants, due to its natural elegance and spectacular orange-scarlet flowers in late spring.

Bodnant Garden is the product of three generations of Aberconways and three generations of Puddles, their head gardeners, a partnership that guaranteed its development and success. Most influential of all the three was Henry Duncan McLaren, 2nd Lord Aberconway, whose great landscaping works raised the status of the garden to one of the greats. Renowned for his passion for plants, in particular rhododendrons, he assembled one of the finest collections in Britain, ensuring that Bodnant in Conway, Wales, remains a "must see" garden today. Because of the glorious surroundings it is a joy to visit, whatever the weather.

Henry Davis Pochin (1824–95), son of a yeoman farmer from Leicestershire, devised the first ever process for making white soap. He used the wealth from this and other successes to purchase Bodnant and its estate of 25 surrounding farms. He also laid the foundations for a remarkable garden.

When he arrived in 1875, there were only lawns, shrubberies, and large trees, many dating to 1792 when the house was built. However, as J. K. Douglas wrote in the *Gardeners' Chronicle* in 1884, "He was an enthusiast in gardening with an accurate

knowledge of trees, shrubs, and all hardy plants." With the help of landscape architect Edward Milner (sometime apprentice of Joseph Paxton), he laid out a large terrace, grassy banks, spreading trees, and the 180 foot long laburnum arch. The garden was "rich in ericaceous plants and had as comprehensive a collection of conifers as was available at the time." His efforts provided the basis of the two-part garden: the formal terraces with their mixed plantings and the rocks and pathways in the wilder river valley below, known as The Dell.

2ND LORD ABERCONWAY VMH

At his death in 1895, Pochin's daughter, Laura, inherited the garden, his love of plants, and the estate. She later married Charles McLaren (1850–1934), who was created Lord Aberconway in 1911.

The garden is mainly the vision of their son Henry Duncan McLaren (1879–1953), the 2nd Lord Aberconway and "one of the most gifted plantsmen of his generation."

From 1904, he designed and supervised the construction of five majestic Italianate terraces, replacing the sloping lawn. Later, he enriched the garden with plants introduced by the great plant hunters, at a time when new plants were prestigious and status symbols. When the famous Veitch nurseries closed in 1914, he bought all their remaining magnolias, commissioning a whole train to deliver them to Bodnant Garden. In 1908 he bought some Ernest Wilson collections, including original seedlings of *Davidia involucrata*, the "handkerchief tree." Frank Kingdon-Ward and George Forrest contributed rhododendrons.

Aberconway's experiences reflected the excitement and uncertainty of plant collecting. In 1926, at the Rhododendron Show in London, featuring the greatest display of rhododendrons ever, George Forrest's introductions won awards. Aberconway was already growing Forrest's Asiatic primulas and rhododendrons from two

previous expeditions, and was eager to add more; he and Fredrick Stern from Highdown (see page 79) met with Forrest in 1928, asking for seed of new alpine plants from "the Lichiang and Tali Ranges" in Yunnan. Aberconway's mood was buoyant; he had just received an award for an exhibition of primulas grown with Frederick Puddle, his head gardener from 1920 to 1947, and had given a well-received paper on "Asiatic Primulas in the Garden." As it was already June, Forrest suggested using his team of loyal, experienced collectors, who knew what to look for without his guidance; he would get in touch with them through his Chinese friends and missionaries. Aberconway and Stern were delighted. Alas, due to communication breakdown, this audacious plan to be ahead of other gardens failed and it was left to Forrest to collect seed the following year. However, Aberconway had also subscribed to Harold Comber's 1926–27 expedition to the Andes, and the plants from this were immediately successful; stands of what is now *Embothrum coccineum* Lanceolatum Group, the red-flowered Chilean fire bush,

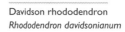

Davidson rhododendron
Rhododendron davidsonianum

The flowers vary in color from soft pink to purple rose and are occasionally spotted. A pink-flowered form from Bodnant Gardens, North Wales, was awarded the AGM (see page 44) in 1935.

HENRY DUNCAN McLAREN

Henry Duncan McLaren was passionate about rhododendrons, the climate in North Wales being ideal for their cultivation. Many gardeners want to grow these magnificent plants but conditions or space don't allow. You can admire them in all their pomp on a grand scale by visiting a great rhododendron garden such as Bodnant. There are, however, many that can be grown on a smaller scale.

→ Rhododendrons are not just about the flowers. Many have attractive young shoots; one of the best is *Rhododendron bureavii*, with pale dun to rusty red shoots. Leaves are beautiful underneath—*Rhododendron* 'Sir Charles Lemon' is cinnamon colored—or they can be beautiful above and below when young, as in *Rhododendron pachysanthum*, which is almost silvery. Others have peeling bark or are fragrant.

Atlas cedar
Cedrus atlantica

There are many forms of this wonderful tree, including what is now known as the Glauca Group, which is widely planted as a specimen tree for its silvery-blue foliage.

→ Rhododendrons have been extensively hybridized and come in a vast array of flower colors, shapes, and sizes. Some of the smaller or slower-growing forms, such as the *yakushimanum* hybrids, are ideal in pots. They should be planted in ericaceous compost and watered with rainwater.

→ Plants grafted on to the Inkarho rootstock, developed from a rhododendron found growing in a German lime quarry, tolerate soils up to pH 7 (neutral), so are suitable for a wider range of gardens.

→ Winter-flowering hardy varieties are a welcome addition to the garden in cool temperate zones, such as *Rhododendron* 'Christmas Cheer', which used to be "forced" for Christmas but can flower in late February, depending on the climate; *Rhododendron* 'Praecox', which is rose-purple in February; and *Rhododendron dauricum* 'Mid-winter', flowering January to March.

still flourish in the garden. A branch from Bodnant, shown in 1947 as *E. lanceolatum* 'Ñorquinco form,' won an RHS Award of Merit. The flower clusters, set close together, "look as if the tree had donned a number of scarlet 'Plus Fours,'" it was noted in the *Journal of the Royal Horticultural Society*.

Aberconway had also persuaded Puddle to try growing Asiatic rhododendrons; Puddle famously doubted they would thrive in North Wales but was happily proved wrong. Puddle had already made a name for himself as an orchid breeder before arriving at Bodnant in 1918. Aberconway and Puddle both began breeding Forrest's rhododendrons in 1920, eventually creating more than 350 hybrids, and also gained widespread success with the fragrant white or flush-pink winter-flowering *Viburnum × bodnantense* 'Dawn', which received an AGM in 1947 and is still popular today. Frederick C. Puddle received a Victoria Medal of Honor for "his work on the hybridization of orchids, rhododendrons, and other plants and for his cultural skill."

In 1939 Aberconway purchased the Pin Mill and removed it from its original site in Gloucestershire, re-erecting it on the Canal Terrace at Bodnant. Legend has it that Puddle persuaded him not to place it in the center of the terrace, where it would spoil the view. And he was right.

The 2nd Lord Aberconway developed and managed the garden for about 50 years, until his death in 1953. He gifted the garden, but not the house, to the National Trust in 1949; it was only the second garden accepted by them on this basis, the first being Hidcote Manor Garden.

Farrer viburnum
Viburnum farreri

A parent of *Viburnum × bodnantense* 'Dawn', it was first introduced by plant explorer William Purdom when he was collecting for Veitch's Nursery.

3RD LORD ABERCONWAY VMH

Charles McLaren (1913–2003), 3rd Lord Aberconway, devoted his energies to the garden, which he managed, extended and enhanced for another 50 years until his death in 2003. He made fortnightly weekend visits to the gardens to admire his rhododendrons, striding round in his plus fours, preceded by an immaculately attired butler. He inherited the presidency of the RHS from his father and became famous for his annual pronouncement: "I think I can say, without fear of contradiction, that this is the finest Chelsea Flower Show ever."

The garden continues to delight and enthrall, mainly due to the dedication of the 2nd Lord Aberconway, his vision, and his love of plants and gardening—particularly his passion for rhododendrons.

Rae Selling Berry

1880–1976

United States

Rhododendron
Rhododendron decorum

One of the most widespread of all Chinese species, first introduced by the plant explorer Ernest Wilson in 1901.

From small beginnings, a few pots by her porch, Rae Selling Berry's collection of native and exotic plants became one of the finest outside major botanic gardens in the USA. From the 1930s, rhododendrons, primulas, and alpines, many of them rare, found their way into her collection. Distinguished botanists and horticulturists became regular visitors to her garden, and she also had many correspondents, all of whom admired her work and discussed and exchanged seeds and plants. After her death, she gave back to nature: her collection of native species contributed to a successful gene bank and conservation program at Portland State University. Her impact as a plantsperson remains to this day.

In 1908, Rae Selling Berry decided to brighten up her porch with a few pots of flowers. From these humble beginnings, her passion for plants grew rapidly. Before long, her garden was crowded with species from the Caucasus, Himalayas, and Alps, and from her own collecting trips in the Pacific Northwest; soon she was renting two vacant lots and they filled up, too. By 1938, there was no more space; so, in her late fifties, she and her husband and most of her garden were transplanted in a new location.

THE NEW GARDEN

The destination, a bowl-shaped, almost 6 acre site near the top of a hill just north of Lake Oswego, in Oregon, had a range of habitats, including springs and streams, a meadow, a ravine, marsh, and wetland, with potential for creating more. A landscape architect designed the plantings around the house and lawn, while Selling Berry developed the remainder of the garden. A true plantsperson, she put the needs of plants first, rather than the aesthetics. Behind the house, log terraces and old stone troughs were a home for primulas and alpines, while raised frames were constructed for those with specialist needs, such as *Rhodohypoxis*.

Her favorites were planted en masse; visitors remember spectacular blocks of bloom. There were beds of *Pleione formosana* (windowsill orchid) or *Shortia uniflora* (Nippon bells). In the 1950s she received a single corm of *Tecophilaea cyanocrocus*, the Chilean blue crocus, at that time believed extinct in the wild; at its peak there were 75 in flower in her garden, a sight almost unknown in cultivation.

INUNDATED WITH SEED

From the beginning, Selling Berry subscribed to gardening magazines, particularly from England, and bought plants from specialist nurseries.

She also discovered that the great plant hunters George Forrest, Frank Kingdon-Ward, Frank Ludlow, and George Sherriff, and the American Joseph Rock, were travelling to China, Burma, and India, searching for plants for cool temperate gardens. Their trips,

sponsored by syndicates, were repaid in seed. "I cannot tell you how eager I am to become a subscriber to the expedition of Mr. Forrest," she wrote. "It's the only way we have the chance of seeing the lovely things, and I do enjoy the game for its own sake." A reply on August 27, 1930 confirmed, "You will be the only one in the USA who has a share in the expedition and, if all goes well, will be in possession of stores of things quite new to American gardens."

Her meticulous attention to detail and keen observation led to a high success rate; her reputation as an exceptional plant propagator soon spread, especially her ability to raise difficult species.

Her favorites were the new primulas; she grew 61 species in all, the largest collection in North America. It was the area of the garden of which she was most proud and on which she lavished most attention.
"I have a nice big bed of Littoniana [now *P. vialii*, the orchid primrose], both the Giant form and the type. This certainly is one of the most striking primulas." (Forrest, 1906)

Chilean blue crocus
Tecophilaea cyanocrocus

One of the most memorable blue flowers in cultivation, this plant is highly prized by gardeners. It is usually grown in pots or in small numbers in rock gardens.

She also grew *P. muscarioides* (Forrest, 1905), *P. sinopurpurea*, "one of Rock's introductions," and *P. florindae* (Kingdon-Ward, 1924) "with huge mopheads of yellow blooms—these flourish with me like weeds,"all named among a seemingly endless list.

She was also believed to have the finest private collection of species of rhododendrons in the USA, numbering more than 2,000 specimens representing 160 species. Dwarf species grew in the rock garden. Two others, *R. decorum* (Forrest, Rock and Kingdon-Ward all made introductions) and *R. calophytum* (Wilson, 1908) were particularly prominent. Large plantings were formed when masses of seedlings rooted through their trays during the Second World War, when there was no labor to plant them. Weak plants died out and strong ones matured, forming a dense canopy of around a dozen stands, 15–19 feet tall, in an area known as the "rhododendron forest," which was spectacular in bloom. Almost all were grown from seed sent by plant explorers, among the rarest a chartreuse form of *R. chryseum* (Forrest, 1918) and *R. sanguineum* ssp. *cloiophorum* (Forrest, *c.*1917).

Passionate gardener, exceptional plantswoman, and an inspirational figure in the world of horticulture, Rae Selling Berry prevailed over a condition of hereditary deafness and left a remarkable legacy for plant lovers.

—Alice Joyce, garden writer

NORTHWEST NATIVES

Selling Berry was also an enthusiastic collector of rare native plants from the mountains of the American West, western Canada, and Alaska. Even in her later years, she could be found in the Wallowa Mountains searching for Oregon's only primrose, *Primula cusickiana* (Cusick's primula)—her "problem child," affectionately known as "Cooky." It defied most attempts to raise it and was short lived in cultivation. From observations on her treks, she became one of the first northwesterners to express alarm at habitat destruction. She grew around 200 of almost 5,000 native plants from that area.

After Selling Berry's death, her native species from the Pacific Northwest became the focus for the first seed bank for endangered native flower species in the USA, dedicated to saving seed and returning plants to the wild. (Laws have changed since Rae Selling Berry's day and it is now illegal to collect plants from the wild.)

Selling Berry was a founding member of the American Primrose Society and the American Rhododendron Society. She also received many honors: an Award of Excellence from the American Rhododendron Society; the Florens de Bevoise Medal of the Garden Club of America, acknowledging "her remarkable knowledge of alpine plants, primulas, and rhododendrons, and success in growing the most difficult subjects;" and a citation from the American Rock Garden Society as one of the great gardeners of America for "the particular study of our native plants."

RAE SELLING BERRY

Rae Selling Berry was not only a great gardener, but a plantswoman extraordinaire. Her success and the size of her collection can be attributed to her overriding passion and enthusiasm for plants and attention to detail. This, and finding the particular area of gardening that you love, can bring pleasure and happiness for years to come. Late each summer afternoon, Rae Selling Berry weeded. Weeds compete with cultivated plants for moisture and nutrients and may harbor pests and diseases, so borders should be kept weed free.

→ Selling Berry grew alpines in raised beds and containers. They are ideal for gardening on a small scale where space is limited, even growing in large pots and windowboxes. This makes it easier to find a habitat in your garden where they thrive, often somewhere protected from scorching sunshine.

→ If you are a beginner, you will find that local garden centers often stock easy-to-grow plants such as sedums and saxifrages and *Erinus alpinus* (the fairy foxglove). As your interest increases, seek out smaller specialist nurseries and alpine or rock garden societies, where a huge range of seeds and plants can be bought and exchanged. You will always receive good free advice from a specialist.

→ Smaller alpine bulbs such as crocus, tulips, and grape hyacinths grow happily in pots. Once they have finished flowering they can be moved and cared for somewhere out of sight, while others that are about to flower take their place.

→ "If you hoe when there are no weeds, you won't get any" is an old gardening saying and it is true, hoeing disrupts germinating seedlings. Selling Berry toured the garden daily to ensure it was well maintained and weed free.

Orchid primrose
Primula vialii

This remarkable plant was originally called *Primula littoniana* and was named for explorer George Forrest's friend, Consul George Litton. It is increasingly rare in the wild.

Sir Frederick Stern

1884–1967

United Kingdom

Winter windflower
Anemone blanda

Sir Fredrick planted a great many spring flowering anemone species in the garden. This one flowered in March, "in all shades of red, pink, and apricot."

Sir Frederick Claude Stern was born in London and educated at Eton and Christ Church College, Oxford. He began a political career as Private Secretary to Lloyd George but, as observed in an obituary, "Luckily for horticulture, this danger was averted." Stern was mentioned with distinction during the First World War, winning the Military Cross; he was briefly an amateur jockey and remained keener on hunting than horticulture until he bought his house, Highdown, near Worthing in West Sussex in 1909. This ignited a love for gardening and plant breeding that remained undiminished throughout his life.

The garden at Highdown, notable because it is created on almost solid chalk, covers just over 8 acres. The house was originally surrounded by small lawns and sheltering windbreaks. The initial focus of Stern's efforts was a disused south-facing chalk pit (originally the site of his tennis court), with its pond and a rock garden designed by Clarence Elliott, a notable alpine nurseryman of the time. As his passion for gardening increased, Stern soon began consulting eminent gardeners of the day, such as Henry John Elwes of Colesbourne Park (one of the first people to receive the Victoria Medal from the RHS), who advised him to "try to raise at least three plants of a species; to put one where one's friends thought it would grow, one where you yourself thought it would grow, and one where no one thought it would grow." A sheer south-facing cliff to one side of the house was so hot and dry in summer that gardeners could only work there in the mornings.

In the early days, a steely nerved garden boy was lowered down on a rope to plant

Cotoneaster horizontalis and Spartium junceum (Spanish broom) along its face. There is barely any soil, yet cotoneaster, whitebeam, and Rosa brunonii, a white-flowered rambling rose, and Arbutus unedo (strawberry tree), an evergreen with attractive bark and strawberrylike fruits, are now well established. These and other plants have been so successful that the cliff is almost covered in foliage.

TRIAL AND ERROR

Stern was constantly experimenting with plants to find out what would grow in such extreme conditions. Most great gardens were growing rhododendrons at the time, and many said it would not be possible to create a garden; however, Stern's trials with magnolias illustrate his tenacity and determination to succeed. He tried over 30 different species until he finally found Magnolia wilsonii, the only one that flourished (the notes on the records in his card index often read "Dead," "dead," "dead!"), and there was great excitement after it was discovered that many Chinese plants, which were being introduced to British gardens for the first time, thrived. He began to subscribe to expeditions led by the great collectors—notably Reginald Farrer's 1914 expedition to the Himalayas and those of Ernest Henry Wilson—so that he could receive seeds direct.

Overall, the most successful plants came from the Mediterranean and China; those from Asia, Japan, and North America were less so. He experimented with gardening techniques, too, and found that plants were more successful in chalk rubble, rather than in holes dug in hard chalk. He thrived on the challenge and wrote, in his typical jocular manner, "It has given me infinite amusement trying to discover what will grow on the chalk, and to make a derelict chalk pit into an oasis." The results of his experiences were written in the most famous of his publications, A Chalk Garden (1960), which is still a definitive reference work on the subject. Stern lists six pages of plants in the index—great encouragement for anyone who thinks that few plants survive in these conditions.

Daffodil
Narcissus cultivars

Daffodils are one of the most popular garden bulbs and a true harbinger of spring. There are so many different kinds that they have been divided into groups to aid classification.

SIR FREDERICK STERN

Stern once noted that a "great number of plants have no dislike of lime and will grow perfectly if the soil is broken up and cultivated." Gardeners today can still apply many of the discoveries he made through trial and error and years of hard work.

→ Bulbs are very successful in chalk, particularly those flowering in early spring. Stern planted masses of daffodils, *Anemone blanda* (winter windflower) with beautiful blue petals, *Anemone pavonia* in shades of purple and red, and species tulips such as the yellow-flowered *Tulipa sylvestris* and red *T. praecox*.

→ Stern discovered that hard chalk had to be broken up to a depth of at least 2 feet for shrubs to survive. This is done by removing the topsoil down to the chalk, breaking through the 4 inch "pan" with a crowbar and mattock, then breaking up the chalk below.

→ Mulching is vital for conserving moisture; flower beds must be mulched with mushroom compost in midwinter, when the soil is still moist. If the soil has dried out, it is too late.

→ Trees successful at Highdown include *Acer griseum* (the "paperbark maple"), noted for its peeling cinnamon-colored bark; *Cercis siliquastrum* (the Judas tree), with bright pink flowers on naked stems in spring; *Sorbus sargentiana*, with sticky winter buds, scarlet fruit, and rich fall color; and *Acer cappadocicum* ssp. *sinicum,* with coppery red spring growth and red seeds in fall.

→ Young plants are more likely to survive when planting in chalk as this gives the roots an early opportunity to establish themselves in the surrounding soil, so the plant develops in response to the conditions.

→ Only tough vigorous species of roses have the vigor to survive in these conditions. Modern hybrid tea and cluster-flowered roses don't have the stamina.

Judas tree
Cercis siliquastrum

This is distinctive because the rosy lilac flowers appear before the leaves in spring. Legend has it that this was the tree in which Judas Iscariot hanged himself.

INSPIRATIONAL PLANTS

The garden at Highdown is at its peak from early spring to early summer, but there is always something of interest, particularly when the roses are in bloom. Among them is *Rosa* 'Highdownensis', which originated in the garden in the 1920s. It is a vigorous, bushy shrub rose, whose new stems often make 10 feet of arching growth in a year and form elegant swags of single crimson flowers and scarlet flagon-shaped hips through the season.

Another of his notable plants is the hugely desirable *Paeonia* (Gansu Group) 'Highdown', with huge snow-white flowers, up to 8 inches across, marked with deep purple blotches in the throat. Peonies, both shrub and herbaceous, were a particular favorite—he wrote *A Study of the Genus Paeonia* (1946) and *Snowdrops and Snowflakes* (1956). He also hybridized many plants, including daffodils and dwarf *Eremurus* species. Several of his plants received the AGM, including *Eremurus* 'Highdown Dwarf' and *E.* 'Sunset'. He was later elected president of the Iris Society and bred hybrid tall bearded iris in the 1930s, such as Iris 'Aline' and 'Marjorie'. Other successes in the garden include hellebores, flowering cherries, and

Himalayan musk rose
Rosa brunonii

This vigorous climbing rose, at its best in a sheltered position or a warm climate, can reach up to 40 feet tall, so it needs plenty of space to climb.

lilies. Most importantly of all, Sir Frederick Stern was the first to prove that highly alkaline conditions were no barrier to creating an interesting, beautiful, and species-rich garden. His garden is now designated a National Collection of the plants he cultivated. Visitors can visit and learn from the results of one of the greatest practical gardening experiments of the 20th century and be reminded that whatever the conditions, with persistence, determination, and by choosing the correct plants, any site has the potential to become a beautiful garden.

If one had to pick out the qualities that made Stern a great gardener, chief among them would be the youthful enthusiasm which he retained all his life and which made him such good company and which was so infectious that a visit to his garden always acted as a tonic to encourage the visitor to renewed experiments.

—The Honorable Lewis Palmer

Jacques Majorelle

1886–1962

Africa

Peruvian apple cactus
Cereus repandus

The flowers, which only open for one night, are followed by edible fruit whose flesh is white, sweet, and full of seeds. It is mainly grown as an ornamental plant.

The painter, craftsman, and gardener Jacques Majorelle was born in 1886 in Nancy, France. His father, Louis, a cabinet-maker, was a central figure in the French Art Nouveau movement, whose designs were influenced by nature, particularly plants. Jacques's studies at the École des Beaux-Arts in Nancy and Paris exposed him to this beauty, too, and he developed a lifelong passion for flora and fauna. He travelled in Spain, Italy, Greece, and Egypt before settling in Morocco in 1919. His garden is an extraordinary fusion of Islamic design, botanical collection, living contemporary painting, and the now famous Majorelle blue.

In 1923, Jacques Majorelle bought a 4 acre plot on the edge of a palm grove in Marrakech and built the house, which he named "Bou Saf Saf" (the Arabic name for poplars), adding adjoining land until he owned about 10 acres. The garden was built on an oasis; he realized this was the only way for a garden to survive in Marrakech. In 1931, he commissioned architect Paul Sinoir to design a second property, an elegant Cubist villa, incorporating his workshop, studio, and living space.

THE CREATION OF PARADISE

From 1924, Majorelle created his garden, with serene reflecting pools, fountains, water channels, plants to create shade, and a diversity of fruit trees (oranges, lemons, pomegranates, dates), reflecting Islamic symbolism and the surrounding gardens. Peacocks, monkeys, Turkish tortoises, and flamingos wandered free among the luxurious vegetation and Majorelle kept two gazelles in a cage in the garden, though the sound of the toads drove him to distraction. "At night the garden was a cacophony of noise," he wrote.

Majorelle loved plants, particularly trees, filling this canvas with rare and beautiful species. He financed plant-hunting expeditions and sometimes traded paintings for cacti; he made frequent collecting trips of his own into the nearby Atlas Mountains, importing and exchanging plants from "all four corners of the earth"—cacti from the American Southwest, waterlilies and *Nelumbo nucifera*, the sacred lotus, from Asia, palms from the South Pacific. Many were introduced to Morocco for the first time. Some provided striking vegetative architecture—sharp angular succulents, spiny barrel-shaped cacti—to contrast with billowing bougainvillea and bamboo, their enmeshed foliage creating texture, subtle change in tone and fragrance, cooling shade, tranquillity, and calm.

Each plant was placed with a painterly eye as, spotlit or backlit, it changed through the day. Majorelle explored the effects of light and shade, giving many of the succulents gallery space so plants could be viewed alone, the space between them becoming a feature. To the artist, it became his "cathedral of shapes and colors," which he often used as a backdrop for his paintings. His passion and desire for plants came at a price, which frustrated and satisfied in equal measure. Majorelle was forced to open the garden to the public in 1847, and wrote of "waiting for an exhibition in Casablanca to replenish my finances which have been gravely depleted by my investment in the garden…you will eventually see that this voracious ogre of a garden, which has been exhausting me for twenty-two years, will need another twenty before it satisfies the angels of Paradise."

Great bougainvillea
Bougainvillea spectabilis

This is one of the many species and hybrids of this famous tropical plant. Its small, insignificant flowers hide within the brightly colored papery bracts.

This garden is a momentous task, to which I give myself entirely. It will take my last years from me and I will fall, exhausted, under its branches, after having given it all my love.

—Jacques Majorelle

MAJORELLE BLUE

In 1937, Majorelle created a bright, intense color, now known as Majorelle blue. He made several expeditions into the depths of Morocco, a colorful country, and would have seen shades of blue being used in buildings in the Atlas Mountains and probably visited Chefchaouen on his travels. With a boldness unheard of at the time in Europe, Majorelle painted the walls of his workshop ultramarine blue, accentuating the greens of the foliage, and adding the outbuildings, gates, pergolas, and pots until he transformed the entire garden. He also added rich red, a color of Marrakech, to the paths and raised beds.

Palms bring a touch of the exotic to any garden, temperate or tropical. Group them together and you will discover that the differences in leaf shape are remarkable.

DEATH, DECAY, AND REGENERATION

Majorelle's later years were overshadowed by tragedy. He suffered a divorce and two serious car accidents. He died in Paris in October 1962 without saying farewell to the garden. As if mourning the death of its creator, it fell into decay.

Yves Saint Laurent and Pierre Bergé discovered the Jardin Majorelle in 1966, during their first stay in Marrakech. They bought La Majorelle in 1980, restoring it and intending to "make the Jardin Majorelle become the most beautiful garden—by respecting the vision of Jacques Majorelle." Gradually, the garden was reclaimed.

In 1998, they approached botanical ecologist Dr. Abderrazak Benchaabane to survey the planting. He returned with suggestions for saving the garden and an inventory of 120 kinds of plants; there are now 325. "I wanted to respect Majorelle's botanical choices so I added new forms from the plant families he chose, enriching them without any shocks, botanical or aesthetic." Automatic irrigation systems were installed and regulated according to the time of day and the needs of each plant, cutting costs and reducing water consumption by 40 per cent.

Jacques Majorelle said, "The painter has the modesty to regard this enclosure of floral verdure as his most beautiful work." He spoke of the garden having "vast splendors whose harmony I have orchestrated;" but it would not be the same without beautiful Majorelle blue.

JACQUES MAJORELLE

Jacques Majorelle took 40 years to create his garden, a labor of love and a feast for the senses. He successfully reflected the surrounding Islamic culture in the structure of the garden, adding exotic plants to create a unique plantsperson's paradise. He freed his imagination.

→ Give specimen plants enough space to develop their natural shape and form. At La Majorelle, plants like cacti and succulents are allowed sufficient space, so each unique shape can be appreciated, rather than being densely planted so they crowd each other out and the lines of their geometry are lost. Growing specimen plants against a plain, contrasting background also increases the impact, be it a succulent against Majorelle blue, or a tree with white blossom with a dark hedge as a background.

→ Strategically placed benches are an invaluable asset to a garden, allowing visitors to rest and admire views, vistas, and focal points and appreciate the surroundings. As a gardener, it is a place to stop, evaluate, and enjoy. It is not just the act of gardening but also enjoying the fruits of your labor that makes it such a pleasure.

→ A garden behind high walls creates its own space and a sense of peace, tranquillity, and isolation away from the hustle and bustle of the outside world (an ever present feature of Marrakech). Tall hedges create a similar effect, but when growing these, always consider your neighbors. Alternatively, create a secluded corner, using plants to muffle the sound.

→ Majorelle added drama to the garden using ultramarine blue; the color is enhanced by the quality of the light in Marrakech and would not have the same impact elsewhere. Paint is a source of instant color but constant repainting is needed to keep it looking "fresh." On a smaller, more manageable scale, add color by painting pots, or buying glazed pots instead.

→ Never underestimate the importance of sound in the garden—frogs croaking, birdsong, the rustle of leaves, the hum of insects, the trickle of water, and the murmur of fountains all add another dimension and soothe the senses.

Prickly pear
Opuntia cochenillifera (L.)

Opuntia, the largest genus among all the cacti, is often planted as an ornamental. They grow extremely rapidly and new plants form when pads break off and root.

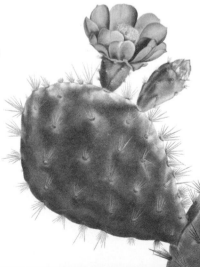

◄ This view of the fountain and rill, with its bold shape, color, and form, encapsulates the spirit of Jacques Majorelle's masterpiece in Marrakech. This is a fine example of turning traditional ideas into contemporary design. The water features, typical of an Islamic garden, respect the garden's context, while the iconic Majorelle blue marks it as the artist's own. Although it is a garden where artistic shape and form are pre-eminent, it is still peaceful, calming, and cool—the quintessential fusion of art and culture.

▼ Pots are perfect for growing plants on terraces and balconies or areas of hard landscaping, and should be used to enhance the design. The choice of plants that can be grown in them is wide ranging. They can be permanent, like trees or topiary, or temporary, like summer bedding plants or tender plants that can go outdoors in summer then be moved indoors in winter in areas of the world where winters are cold. Pots must be large enough to accommodate the roots, and stable, so the wind doesn't blow them over. You can grow almost anything in a pot, providing there are drainage holes in the base. If there are no drainage holes in the pot, grow bog plants or, if it is large enough, create a miniature pond.

Madame Ganna Walska

1887–1984

United States

Dwarf water lily
Nymphaea candida

Ganna loved exotic and unusual plants so this dwarf hardy waterlily would have almost certainly appealed. With flowers only 2 inches across, it is ideal for a small container.

Ganna Walska bought an estate in California, created a series of gardens and called her astonishing creation "Lotusland." It was gardening on an impressive scale because of the volume of plants that she used, many never seen before in such numbers. Although some of her work was inspired by other cultures, her unique creativity added to the glamor. Her own gardens were pure theater, at once bizarre, beautiful, awe-inspiring, and unique.

Walska's extraordinary life explains how she financed her extravagances. Born Hanna Puacz in Poland, probably in 1887 (her 10 passports show different birthdays), she eloped with a Russian count at the age of 20. She then began a career as an opera singer and found her stage name: Ganna (Hanna = grace). She loved dancing; Walska is a derivative of Waltz. She thus became known to the world as Madame Graceful Waltz.

Her beauty, magnetism, and personality were a bait that caught six husbands, all but one extremely wealthy, including Alexander Cochran, known at the time as the "richest bachelor in the world." She managed to keep lifelong allowances, a home in Manhattan, another in Paris, along with the Théâtre des Champs-Elysées, and a chateau in Galluis, Ile-de-France. After an unsuccessful operatic career she turned to Eastern mysticism, and met and married Theos Bernard, the self-proclaimed "White Lama," over 20 years her junior. In 1941, he found an estate in Montecito, north of Los Angeles, California, and persuaded her to buy it; four years later, they divorced. It was after this turbulent period that she was to turn her attention more fully to her other passion in life: gardening.

Lotusland

Husband free, Walska lavished the remaining years of her life on her own 37 acre operatic production, described by Sean K. MacPherson in the *New York Times* magazine as "the living incarnation of Walska's irrational exuberance." She filled the garden with rare and exotic species, falling in love with their beauty, but never knowing their names. Wanting plants that always looked good, Madame (she called herself "head gardener," but only answered to Madame) was more interested in their art, style, and beauty. Before creating each garden, she compiled scrapbooks of articles and photos. "It seems to me the best way to develop creativeness is to learn the techniques of the great artists and to imitate them. First to do as well as they do, then to try and surpass them; then to surpass yourself."

The garden

Walska was helped in its construction by Santa Barbara's most famous landscape architect, Lockwood de Forest (though she complained that all he did was plant rocks) and a team of gardeners and contractors. She created 18 different areas. There are gardens for butterflies, cacti, and ferns; an Australian and a Japanese garden; a topiary garden (with 26 topiary animals, including a camel, a giraffe, and a seal);

plus several bizarre flourishes, such as a topiary horse with light-bulb eyes and a magnetic rock onto which Walska would drop her hairpin and it would stick—all in her own inimitable style.

Outside the front door of her pink stucco house is a grove of *Dracaena draco* (dragon trees), some dating back to the 19th century. The story goes that she added to the collection by driving around the neighborhood. When she found a good specimen, she had her chauffeur knock on the front door, offering to buy them. If they didn't want to sell, a case of champagne was delivered; that usually changed their minds!

Glaucous echeveria
Echiveria secunda var. *glauca*

In cool temperate climates echeverias are grown as houseplants or used in outdoor bedding schemes; in Mediterranean conditions, they make an excellent year-round display.

MADAME GANNA WALSKA

The gardens and the hugely diverse range of plants at Lotusland are now being carefully curated and managed sustainably. It is living proof that a garden can be an ornamental extravaganza and an example of good horticultural practice at the same time.

→ At Lotusland, there is a butterfly garden. A garden is not just for you but for nature, providing nectar, egg-laying sites, and even caterpillar food. There are plenty of books and websites to help you to decide what to plant.

→ The garden has its own well. It is worth checking records, particularly if you live in an old house, to see if there is a well; if restored, it will be a massive advantage to your garden. Before using it, do check with the local authority about laws relating to abstraction.

→ The colossal plant collection at Lotusland now numbers 208 different plant families and 3,286 taxa or different kinds of plants, including cultivars. Many were and are very rare but their future is secure in the garden; what was once an indulgence has become an invaluable scientific collection. Practice back-garden conservation by growing a few rare wild

or cultivated plants of your own and join the societies that support this venture.

→ Despite the wide collection of plants the garden is managed sustainably, with no chemical fertilizers or herbicides. You can do that too.

→ Walska created scrapbooks of articles and photos before making her garden; these were her "mood boards," which many landscapers use. A mood board allows you to get a feel of the garden you are creating before you start buying and planting.

→ Follow your own style. Many are influenced by the "greats;" why not be original?

Queen of the night
Epiphyllum oxypetalum

The flowers of this cactus are exquisitely beautiful and deliciously fragrant. Individual flowers only last for one night. Grow in a frost-free conservatory or glasshouse in cooler climates and enjoy!

Lotus garden

From 1953 to 1956, Walska almost single-handedly oversaw the conversion of the original swimming pool into a water garden, which she filled with her favorite lotuses, and then built a new pool for bathing. In 1958, she collaborated with local artist Joseph Knowles, Sr., constructing the spikily beautiful aloe garden and decorating the edge of the kidney-shaped reflecting pool with abalone shells; on two coral plinths, yawning clamshells spill water from their mouths. There are more than 170 different aloes in this garden.

Most would be satisfied with individual specimens but Walska filled her beds with dozens, sometimes hundreds, of the same species, coming up with the idea of mass plantings. Her philosophy seemed to be "if one is good, a hundred is better." When the Palm Society was due to visit; she panicked, thinking she didn't have enough, sold some jewellery, bought more palms and ended up with 375.

Her concern with effect is amply demonstrated in the iconic blue garden, featuring plants with silvery to glaucous leaves that look ethereal in the moonlight, including *Cedrus atlantica* 'Glauca' (blue Atlas cedar), *Festuca ovina* var. *glauca* (blue fescue) and *Brahea armata* (Mexican blue palm). Chunks of blue-green glass, chipped from the kilns of a water-bottling factory, line the pathways. Only Walska would see these as costume jewellery for a garden.

Winterberg cycad
Encephalartos cycadifolius

Cycad garden

The cycad garden, featuring over 200 species, was Walska's last waltz. She became less and less of a socialite as it grew, selling almost $1 million worth of jewellery to pay for the pleasure. There are over 900 specimens, representing half the known species, one of the world's finest collections. The rarest is *Encephalartos woodii* (Wood's cycad). No female plants have ever been found; it is extinct in the wild and one of the most sought-after cycads in the world. A collector would be ecstatic to have one; Walska had three.

Until the last days before her death in 1984 at the age of 97, Walska walked the garden daily with the help of two walking canes. "The word 'impossibility' does not exist in my vocabulary any more. Nothing is impossible!" she said, also admitting, "Never could I do anything comfortably or halfway" and "I have a particular aversion to following the multitude in styles of any kind."

Walska, "an enemy of the average," truly vanquished her foe, creating a garden of distinctive and extraordinary genius.

I'm an enemy of the average.

—Madame Ganna Walska

The fern garden, designed by landscaper William Paylen, was extended several times before being completed in 1988 and was built around Madame Walska's collection of Australian Tree ferns (*Cyathea cooperi*) with their elegant, plumelike fronds. There are now many different types of tree ferns in the garden, species and cultivars, all under-planted with terrestrial ferns (there are even giant stagshorn ferns, *Platycerium* species, growing on the trees). The whole garden in myriad shades of green is punctuated with shade-loving begonias, calla lilies (*Zantedeschia* cultivars), and clivias (*Clivia* species and cultivars) to provide a welcome splash of color.

Madame Walska, who was influenced by Eastern religions, named her home Lotusland after the beautiful water lily, *Nelumbo nucifera*, and transformed the original swimming pool into a lotus pond, growing several different varieties. The sacred, or divine, lotus is widely planted in Mediterranean and tropical climes for its beauty, fragrance, and religious significance. Whether its religious symbolism is significant to the individual or not, there is no doubt that it is an exquisitely beautiful, graceful garden flower.

Vita Sackville-West

1892–1962

United Kingdom

Filbert
Corylus maxima

This grows into a large shrub or small spreading tree, with large edible nuts. Many ground cover plants in the garden thrive happily in its shade.

Born into an aristocratic family at Knole, Kent, in 1892, Vita Sackville-West went on to create gardens in Constantinople and Sevenoaks before transforming the derelict grounds of a castle at Sissinghurst into one of the most famous gardens in the world. Her diplomat husband Harold Nicolson liked the rational and classical; she, the poetic and romantic. A poet (one of her most famous works is entitled "The Garden"), novelist, and radio broadcaster, Sackville-West also wrote a popular weekly gardening article in the *Observer* newspaper from 1945 to 1961, which was later published in several volumes. In 1955, she received a Gold Veitch Memorial Medal from the RHS.

One spring day in 1930, Vita Sackville-West visited a property in the village of Sissinghurst in Kent. The estate agent's description was of a "farmhouse with some picturesque ruins in the grounds." "I fell in love; love at first sight. I saw what might be made of it," she wrote.

PLANNING AND PLANTING

Work began in 1930, almost as soon as they moved in. Nicolson, a descendant of the 18th-century architect Robert Adam,

designed the layout, dividing the plot into interconnecting gardens "like the rooms of an enormous house," a "succession of privacies" around the Tower, using strategically placed walls, hedging, paths, and vistas. As he was working abroad at the time, most of the consultation was by letter. Garden rooms draw visitors further into the garden, fuelled by curiosity and laden with expectation. The rondel (an old Kentish word for the hop-drying floor of oast houses), a circular yew hedge with an immaculate central lawn with four paths leading from the quarters, is a perfect example. Their friend Edwin Lutyens helped devise a plan for a small parterre of L-shaped beds.

Sackville-West painted the rooms with plants, each division with a concentration of flowers from a specific season or a color scheme; sometimes with bold, extravagantly colored brushstrokes, at other times soft and subtle; but always with impeccable taste. The style was traditional with exotic Mediterranean touches, Sackville-West's passion for plants influenced by her love of old Dutch flower paintings and the plants she saw on her travels.

This harmonious collaboration between Nicolson and Sackville-West created a garden combining "the strictest formality of design, with the maximum formality in planting."

SOME FAMOUS ROOMS

Plantings in the courtyard use salmon and copper tones—roses such as the richly fragrant, buff-yellow-to-pink climber 'Gloire de Dijon' and the salmon pink rambler 'Paul Transon.' To one side, there is a purple border (a color disliked by their friend Gertrude Jekyll), which, when it was

planted in 1959, had a limited range of shades. Pamela Schwerdt and Sibylle Kreutzberger (joint head gardeners 1959–90) increased that range, adding lighter colors to prevent the border from becoming too dark while avoiding consistent gradation in borders, which Sackville-West disliked. Among them were scrambling *Geranium psilostemon*, with cerise flowers and striking black centers, indigo-flowered *Clematis × durandii*, and dark purple *Cotinus coggygria* 'Foliis Purpureis,' a form of smoke bush.

Of all the plants at Sissinghurst, roses captured Sackville-West's imagination the most, starting in early summer, her favorite time of year. They thrived in the rich soil of what was once the kitchen garden. Her favorites included *Rosa* 'Tuscany Superb,' with crimson petals and a boss of golden stamens; deep purple *R.* 'Cardinal de Richelieu,' one of the darkest-flowered roses; and large-flowered *R.* 'Tuscany.' The ground below is a carpet of traditional herbaceous plants—irises, peonies, and pinks—while the curved wall at its west end is draped in a curtain of free-flowering, pale blue *Clematis* 'Perle d'Azur,' to dramatic effect.

The Nuttery already existed when the couple bought the property. Nicolson added foxgloves, collected from the surrounding woods in an ancient pram; Sackville-West added polyanthus in a range of colors, which, despite intensive care, eventually died out. A new color scheme of yellowy-green, blue, and

Tuscany superb
Rosa gallica

There are two varieties: 'Tuscany' and 'Tuscany Superb'. *Rosa* 'Tuscany Superb' (a sport of 'Tuscany'), known since 1848, is more vigorous, deeper red, and has better foliage than its parent. Sackville-West particularly loved 'Tuscany', the 'Old Velvet Rose'.

white was planted in 1975, long after
Sackville-West's death. Now spring is
heralded by the unfurling elegance of
Matteuccia struthiopteris, the shuttlecock
fern; aristocratic *Trillium grandiflorum*, a
wake-robin with simple white blooms; and
a tapestry of other shade lovers, all embraced
by a canopy of arching hazel.

The Orchard is swathed in climbing
roses with pools of narcissus and wild flowers
below, and the Cottage Garden is a vibrant,
warming palette of sunset oranges, reds,
and yellows, from tulips and wallflowers
in spring, to tender subjects such as
Hedychiums, salvias, dahlias, and cannas
in late summer and fall. "I used to call it the
sunset garden in my own mind before
I even planted it up," she wrote.

She walks in the loveliness she made,
Between the apple-blossom and the water—
She walks among the patterned pied brocade,
Each flower her son, and every tree her daughter.

—*The Land*, Vita Sackville-West

But whatever has gone before, the White
Garden, much copied but never bettered, is
the most famous of them all. It is a garden
of romance, subtle in flower and foliage and
a mass of seductive fragrances. In her first
article on her idea of creating a Gray,
Green, and White Garden (the *Observer*,
January 22, 1950), Sackville-West wrote: "I
cannot help hoping that the great ghostly
barn-owl will sweep silently across a pale
garden, next summer, in the twilight—the
pale garden that I am now planting, under
the first flakes of snow." By 1954, most of
the main plants were in place. In spring,
white bells of *Polygonatum* combine with
the tumbling foliage of a silver weeping
pear; by midsummer, a cloud of single-flow-
ered *Rosa mulliganii* billows from the mock
gothic arbor and white flowers abound—
among the highlights are the elegant spikes
of *Chamaenerion angustifolium* 'Album'
(white rosebay willowherb); fragrant *Lilium
regale* (regal lily); and *Onopordum acanthium* (Scots thistle).

It is the garden's grand finale.
Sackville-West wrote to Nicolson in
1962, just before she died, "We
have done our best and created
a garden where none was." It is
so much more than a garden.

VITA SACKVILLE-WEST

Vita Sackville-West was an imaginative, skilful gardener who proved the old adage: "To know it is to grow it." Her work was testament to the fact that a garden never stands still and you should always seek ways in which it can be improved.

→ At the start of the Second World War, when a German invasion of Britain was anticipated, Vita Sackville-West planted 11,000 daffodils as a lasting memorial should they have to flee. Naturalizing bulbs (that is, planting in a seemingly natural manner) is a simple, effective way to create an impact, even on a smaller scale; but do not mow the grass until after the leaves of the bulbs have died back in spring.

→ Vita removed any plant if it was not exactly right but disliked excessive tidiness, encouraging self-seeding and wild flowers. "Nature sometimes betrays a divine instinct for shedding her nurslings in the right place," she wrote. It is remarkable how serendipity plays a part; leave self-seeded plants, and don't remove them until you are certain they are going to have a negative impact.

→ One year, Vita returned from a winter cruise and was delighted with how her head gardener, Jack Vass, had trained the roses. To her, it evoked memories of "a Cornish harbor" with the shrub roses tied down over hoops looking like "lobster pots." This technique is also adaptable for climbing roses: bend the shoots down in half circles, fastening them individually into the wall or on to the shoot below. Doing so produces masses of flowers, avoiding a bare base.

→ Successive head gardeners have revised the planting schemes at Sissinghurst in the spirit of the garden, replacing old plants with newer, similar, and better varieties, extending the season of interest for visitors.

→ There is little use of bold foliage and contrasting form, variegated, or punctuation plants. The White Garden proves that single colors can be used effectively with subtle changes in tone.

→ Choose white-flowered plants carefully; their blooms turn brown when they die unless deadheaded, or the flowers are small, diminishing the overall impact.

Solomon's seal
Polygonatum odoratum

Polygonatum odoratum in its variegated form has been planted in The Nuttery. It was added in later years as part of the continuing development of the garden in the spirit of Vita Sackville-West.

The twilight at dawn and dusk or clear moonlit summer nights reveal Vita Sackville-West's White Garden at Sissinghurst at its most romantic. It is filled with different shades of gray, green, white, and silver, with further interest added by introducing different textures, shapes, and forms. You may not have space to create a large-scale homage to this influential iconic garden, but you can plant up borders or containers—these cool, sophisticated colors are particularly useful for brightening shady corners. The garden is laid out in a formal design and the brickwork in the paths is the same pattern that was used by Beatrix Farrand at Dumbarton Oaks (see page 96).

Margery Fish

1892–1969

United Kingdom

Buxton's variety
Geranium wallichianum

Herbaceous geraniums are robust, long flowering, and beautiful. This is a low growing form with long, running stems.

Margery Fish was patient and hard-working—useful virtues during 20 years in Fleet Street and later as a gardener. When Lord Northcliffe was asked by Prime Minister Lloyd George to head a mission to the USA in 1917, he requested that Margery be his PA. Undeterred at the thought of crossing the Atlantic under threat of enemy torpedoes, she accepted and was awarded an MBE for her contribution to the war effort. During her career, she was secretary to six *Daily Mail* editors, including Walter Fish, whom she married three years after he retired. Then she made a garden.

In 1937, with war imminent, Margery and Walter Fish decided to move to the country. While house hunting, they made a brief visit to dilapidated East Lambrook Manor, South Petherton, Somerset, with its 2 acre rectangular garden, "But it was such a wreck that Walter refused to go further than the hall." After three months of fruitless searching, they returned and bought the "poor battered old house that had to be gutted to be livable, and a wilderness instead of a garden," she wrote. Their friends wondered how two Londoners would go about the job of creating a garden from a farmyard and rubbish heap.

MAKING A PLAN

In her late forties, without any knowledge or ever having shown the slightest interest in gardening, Fish set to work.

He wanted rolled gravel paths, neatly edged lawns, giant delphiniums, dahlias, and hybrid tea roses; she wanted an abundance of cottage-garden flowers, self-sown seedlings, and interest all year round. He thought he was in charge; she thought she was in charge. However, both agreed that "a good bone structure must come first, with an intelligent use of evergreen plants so that the garden is always clothed, no matter what time of year…a good garden, is the garden you enjoy looking at even in the depths of winter."

She started by the house, with the Terrace Garden. "I studied the ground for days on end, looked at it from every angle, drew plans on paper, and by degrees, ideas took shape." Fish's energy was legendary; remembered as "fast-talking and quick-moving," she managed the garden with only occasional help, often working 18-hour days, even in her senior years, writing books and articles early and late, as she did while at Associated Newspapers.

Over time she dug trenches, built drystone walls, laid winding narrow paths, constructed terraces, and barrowed wet clay along slippery planks in winter, creating a garden that incorporated several areas. "One of the things we tried to do was make the garden as much part of the house as possible…so we made the garden round the house…the hall was paved with flagstones and we paved the garden outside…it was difficult to tell where one ended and the other began."

Primrose (various)
Primulaceae

Primroses, cowslips, and others are happy to naturalize, in the carefree spirit of cottage gardening. Like many plants, they find the right spot to suit their needs.

The garden was as "modest and unpretentious as the house, with crooked paths and unexpected corners," just as she intended.

Fish was also a formidable plantswoman, mixing old-fashioned everyday plants with rarer species. Shady corners were packed with primroses and epimediums, and herbaceous plants, and shrubs filled the borders; and there was always some small delight to be found nestling near the stream or falling over the stone paths and walls. "It is pleasant to know each one of your plants intimately because you have chosen and planted every one of them."

The Silver Garden, traversed by a meandering path catching the heat of the day, was home to artemesias, dianthus, and spiky *Cynara cardunculus* (cardoon, or globe artichoke); a damp, shady garden utilized the stream that ran behind the old malthouse and in the Ditch she grew moisture-loving plants and snowdrops and hellebores—some of her favorites. The garden was informal, comfortable, and carefree, with understated, unfussy planting ideas, to which anyone could adapt and relate.

Fish bought plants from local nurseries and by mail order, swapped cuttings with gardening friends, and made selections from plants she discovered in the garden; *Artemisia absinthium* 'Lambrook Silver' (wormwood), *Euphorbia characias* ssp. *wulfenii* 'Lambrook Gold' (spurge) and *Primula* 'Lambrook Mauve' were her introductions. She promoted unfashionable plants such as hostas and bergenias and loved herbaceous geraniums. "I find them a most adaptable and generous family. Many of them flower throughout the season, they fit in anywhere and in many cases good foliage is another recommendation. I wouldn't like to be without any of them in my garden." In the late 1950s she opened a successful nursery, which became famous for its wide range of plants; visitors could then take something of East Lambrook home.

GOOD PLUS BAD EQUALS EXPERIENCE

As a writer, Fish's approach was accessible, honest, and refreshing. In *We Made a Garden* (1956), she was candid enough to include a chapter "We Made Mistakes"—a rare admission for someone who had become an expert. Her advice is practical, varied, sometimes novel, and borne of personal experience. "In winter there is always a wood fire smouldering on the open hearth in the hall and it makes a wonderful pyre for dangerous weeds… and a convenient source…when potash or charcoal are required for garden operations."

Fish wrote seven more books and contributed to several others. In the late 1950s she had a regular column in *Amateur Gardening* and then in *Popular Gardening*, and appeared regularly on the BBC. When a database of every plant she mentioned was compiled by the nursery at Lambrook in the 1990s, it ran to 6,500 names and more than 200 single snowdrop varieties.

Fish was awarded a Silver Veitch Memorial Medal by the RHS in 1963 for creating her garden and her writing. After she died her nephew and later owners developed the garden; through it her joyous, carefree, friendly spirit remains with us today.

Species hellebore
Helleborus dumetorum

Hellebores can be bold and blowsy, or small, dainty, and elegant; this plant is the latter, with a quiet, serene charm. Subspecies *atrorubens* has purple flowers.

We all have a lot to learn and in every new garden there is a chance of finding inspiration—new flowers, different arrangement, or fresh treatment for old subjects. Even if it is a garden you know by heart there are twelve months in the year and every month means a different garden, and the discovery of things unexpected all the rest of the year.
—Margery Fish

MARGERY FISH

Margery Fish's books are filled with practical advice; they are simple, conversational, and easy to understand. Her husband, Walter, had some interesting ideas and observations and knew more about gardening than he would often let on!

→ Use old barrels to collect rainwater to water your plants; they are more attractive than plastic butts. Water can be drained from the guttering via a downpipe or down a heavy-duty chain.

→ "Alpine plants were tucked into the cracks and crevices when building drystone walls…white arabis and stonecrops, with some rock campanula."
This is an ideal way to turn drystone walls into a feature. Select plants according to their needs—sun or shade—and plant them in niches of suitable compost, adding grit where necessary.

→ "The four essentials of a good garden are perfect lawns, paths, hedges, and walls; if the surroundings are unkempt, the flowers will give no pleasure… Walter would no more have left his grass uncut or the edges untrimmed than he would have neglected to shave."

→ "I could always count on a harvest of choice little seedlings in the (gravel) drive." Plants self-seed freely in gravel; let them remain or carefully lift and transplant them elsewhere.

→ Deadheading—the removal of spent flowerheads of any plant except those that produce ornamental fruit or from which you intend to collect seed—ensures energy is not wasted. "I even deadhead my naturalized daffodils. I have some old swords and I keep one sharpened for this job. One can slash off a lot of heads in a very short time."

→ Firm planting is one of the first essentials and it is a good idea to give a little tug to anything that is just put in to make sure it is firmly anchored.

Large masterwort
Astrantia maxima

A vigorous, spreading plant that needs rich moist soil and good light to perform to its potential; when it does, it is an outstanding plant.

Beatrix Havergal

1901–1980

United Kingdom

Royal Sovereign strawberry
Fragaria x *ananassa* 'Royal Sovereign'

This mid-season variety bred by fruit breeder Thomas Laxton dates back to 1892. What they lack in size, they more than make up for in flavor.

Beatrix Havergal, gardener and educator, is remembered for striving for perfection in horticulture and being a pioneer in the education of women at her center of excellence, Waterperry Horticultural School for Women in Oxfordshire. A brilliant practical gardener and natural teacher, she delighted in sharing her enthusiasm and knowledge with others. She was also formidably determined and strict, yet with her warm personality, keen sense of humor, and zest for life, Beatrix Havergal inspired loyalty among her students and staff and was greatly admired and respected throughout the horticultural world.

Havergal's father was a vicar and her early life was spent on the move between places as diverse as Norfolk, Worcestershire, Paris, and a boarding school in Kent. She left school in 1916 and took gardening jobs under the auspices of the Women's War Agricultural Executive Committee, tasked with increasing agricultural production in every county of the United Kingdom in the First and Second World Wars. Trix, as she was known to the family, had a beautiful contralto voice and was a highly accomplished cello player, but the cost of musical training was too high. Gardening seemed a safer bet for finding work and so she chose to study horticulture. She went to Thatcham Fruit and Flower Farm near Newbury, Berkshire, where women were trained in the art and craft of gardening, graduating with an RHS Certificate with Honors. Her first task was to design and build a garden at Cold Ash, near Newbury, where the quality of her work was noticed by Miss Willis, headmistress of nearby Downe House boarding school, who invited her to become head gardener.

Havergal longed to teach as well as to garden, an idea that would later lead to her starting her own school. At Downe House she met Avice Sanders, the housekeeper, who became her lifelong companion. In 1927, with the blessing of Miss Willis, and capital amounting to less than £250, Havergal and Sanders moved to a cottage with 2 acres of walled gardens in the grounds of Pusey House, near Faringdon, Oxfordshire, and took their first students. Short of money, they grew crops to sell at Swindon market. The course aimed to combine theory and practice with high standards of speed and efficiency—the graduates had ample job opportunities. Havergal also continued her studies and in 1932 obtained the RHS's highest qualification, the National Diploma of Horticulture (now RHS Master of Horticulture).

Delphinium
Delphinium cashmerianum

There are stately perennial delphiniums for herbaceous borders, dainty annuals for cottage gardens, and others, like this Himalayan species, that thrive in rock gardens.

On the move (again)

The school was so successful that in 1932 they moved to larger premises at Waterperry House, near Oxford, which they rented from Magdalen College, spending the first five years clearing trees, building glasshouses, and improving the soil for production. By the outbreak of the Second World War, the school offered a residential two-year course for 15–20 students, from any background, from Britain or abroad. At the time, students were self-funded, but the school later became recognized by the Department of Education and scholarships were granted by some county councils.

The syllabus encompassed every aspect of gardening. The 1937 prospectus stated: "It is the object of the school to provide a theoretical foundation, the practical knowledge of Horticulture, and the practical skill to make a first class gardener." Havergal was a great communicator, knowledgeable, theatrical, and fun. Each student was responsible for a section of glasshouse, taking turns to stoke the boilers, and tasks were organized so they became skilled all-rounders. Students and staff at Waterperry worked together and the gardens were gradually developed to become teaching units: fruit, flowers, and vegetables were grown and new glasshouses were constructed. Some of the crops they produced were sold at Oxford market. Towards the end of their two-year course, students took the RHS General Examination and the Waterperry Diploma. Havergal fought hard for it to be given equal standing with the equivalent male qualification. By 1960 the Waterperry Diploma was accepted by the Institute of Park Administration and regarded as being on the same level as the Kew and Edinburgh diplomas. Although it was a horticultural qualification, punctuality, reliability, friendliness, and willingness were also taken into account. Many great gardeners were influenced by Beatrix Havergal, including Pamela Schwerdt and Sibylle Kreutzberger, who were head gardeners at Sissinghurst (see page 123); Vita Sackville-West referred to them as "her treasures, so well trained."

Havergal was a fruit enthusiast but also loved the herbaceous border she designed, 200 feet long and backed by the red brick nursery wall; with flowers from late May until the first frosts, it starts with plants such as lupins, geraniums, and veronica, followed by delphiniums, verbascum, and phlox, and concludes with late herbaceous plants such as heleniums, rudbeckias, and asters. During the war, the two-year course was replaced by short courses for women in the Land Army, and a further 30 acres were turned into a market garden as part of the war effort. From 1943, Havergal's "Dig for Victory" demonstrations were held, teaching people how to produce their own fruit and vegetables; they were continued after the war in a bid to reach a wider audience. In 1948, Havergal and Sanders were finally able to buy the estate with the assistance of an anonymous benefactor and, by 1963, day classes were started.

Beatrix Havergal was appointed MBE on February 23, 1960, and later the same day was awarded the RHS Gold Veitch Memorial Medal. In 1966, she was awarded the Victoria Medal of Honor; she was also president of the Horticultural Education Association.

Havergal was admired throughout the horticultural world and was well known to a wider audience for her exhibits of strawberries at the Chelsea Flower Show, where she received 15 gold medals over the course of 16 years. Havergal continued to play the cello and to sing with the Newbury Choral Society and later with the Oxford Bach Choir. She also joined the Thatcham Special Police Rifle Club and became a skilled shot. In 1971, with her health deteriorating, Havergal sold the estate and retired to a cottage in the grounds. She died on April 8, 1980 and was buried in Waterperry churchyard.

Her success as a teacher was due to her warm, enthusiastic personality and high standards. She made you feel that only the best was good enough and you could achieve it. To do so it was vital to understand what you were doing, why you were doing it, to see it done correctly, and then do it yourself. You would then be proud of your work and know enough to "cut corners" and still be successful.

—Mary Spiller, friend of Beatrix Havergal and BBC *Gardeners' World*'s first female presenter

BEATRIX HAVERGAL

Beatrix Havergal was a craftsperson par excellence and students could easily learn a technique by watching her do it correctly. Something of an actress, she would often overstate to emphasize a point. She wrote: "The only way to learn is to do the work." Gardening is not only a task, it is also a craft. Below are some of her gardening tips.

→ Level ground and even spacing ensures uniform crops of fruit, flowers, and vegetables. It also looks professional.

→ Havergal always had the first day of March as her deadline for completing the winter work, so the garden was prepared and ready for the new season. Finishing your winter tasks with time in hand is always a bonus; there is so much to do in spring.

→ Use a straight-edged piece of wood, calibrated with plant spacings, to make your rows in the vegetable garden. Straight lines save space, are pleasing to the eye, make cultivation quicker, and you are less likely to hoe off plants accidentally.

→ The old saying "The answer lies in the soil" is correct. Improve the structure with well-rotted organic matter and avoid walking on the soil, particularly when it is wet, because this causes compaction and breaks down the structure.

→ Hoe frequently; a loose surface makes this job easy and quick. It also kills the weeds when they germinate or while they are still seedlings so hand-weeding is not necessary.

→ Stake plants early in the year, well before they flop, pushing the stakes deep enough into the ground that they are able to support the weight of the plant.

→ Prune carefully to maintain the natural shape of any woody plant, tree, or shrub and also to encourage flowering. Some plants flower on the current year's growth, others on growth made the previous year, so make sure you know how to prune correctly or flowering wood may be lost. If in doubt, think how it reacted to last year's treatment.

→ When picking apples, cup your hand underneath the fruit, lift and twist. Don't use your fingertips; if you can feel pressure on them, you are picking incorrectly.

Cox's orange pippin apple
Malus domestica 'Cox's Orange Pippin'

Originating in 1830, it is regarded by many as the best of all eating apples; juicy and crisp, aromatic, spicy, and honeyed, it has a complex flavor. It needs warm conditions.

Following the principles of planting and color proposed by Gertrude Jekyll, Havergal created a border purely from herbaceous plants at Waterperry Gardens. There are three peaks in the flowering season. May and June bring plants like lupins, herbaceous geraniums, and veronica. Their fading flowers are hidden by others like delphiniums, phlox, and achilla, timed to peak on July 7, Miss Havergal's birthday. Finally, heleniums, Michaelmas daisies, and goldenrod take center stage from September until the first frosts. To extend the season of interest in late winter and spring, plants like pulmonaria, snowdrops, and alliums have now been added.

The formal gardens, with their surrounding yew hedge, were the brainchild of Mary Spiller and Bernard Saunders, an artist and steward of Waterperry Gardens, and were created to acknowledge Waterperry's 500 years of history. A Tudor-style knot garden, with traditional clipped box hedging and topiary cones, the centerpiece of the design, is bedded out or planted with herbs according to the season. The purple, clipped shrub is *Berberis thunbergii* f. *atropurpurea* 'Atropurpurea Nana'. In borders along each side of the garden are plants representing historical periods from Tudor times until the present day.

Mildred Blandy

1905–1984

Portugal

King protea
Protea cynaroides

Proteas are trophy plants for any gardener, particularly if they can be grown outdoors. The king protea has the largest flower of the genus.

Mildred Blandy was born to an English father and Irish mother in East London, South Africa. In 1930, aged 25, she sailed from Cape Town to Madeira on the maiden voyage of the liner RMMV *Winchester Castle* to visit relations and met Graham Blandy. They married in London in 1934 and settled at the family home, Quinta do Palheiro in Funchal, Portugal. Blandy took a keen interest in the gardens from the start, importing plant material from around the globe. In doing so, she created one of the most celebrated gardens on the "Island of Flowers."

The garden at Quinta do Palheiro dates back to 1801, when the first Conde de Carvalhal bought the estate. He employed a French landscape architect to lay out the Quinta, as evidenced by the grand avenues of plane and oak trees. He also imported specimen trees from around the world to fill the estate. Tradition says that he was gifted many rare species, such as *Araucaria angustifolia*, by the King of Portugal, Dom Joao VI, and others from masters of ships visiting Funchal harbor. He also began the camellia collection, importing plants from Portugal and Belgium. An Englishman who visited on January 13, 1826 wrote, "The camellias are the principal ornament producing red and white flowers which rival the rose in form and color but do not have the beautiful scent."

After the bachelor count's untimely death, the second count, his nephew, squandered the inheritance. In 1885, English merchant John Burden Blandy bought the estate, built a new house, and added a winding avenue with camellias on one side, with a range of species and cultivars flowering from the beginning of winter to early spring.

A life's work

Although the garden has been developed by successive generations of "Blandy women," it was Mildred Blandy, a generous, kind, yet rather shy lady, who gave the gardens their character. Happiest when in the garden checking on her new plantings, she cried when her dogs ravaged her new planting, collected seeds and cuttings from wherever she visited, and never stopped thinking about her glorious garden. She gardened these 30 acres from 1935 to 1984, making the garden famous.

The garden's style is very much late Victorian but bejewelled with surprises. There is a sunken garden, herbaceous borders filled with tender plants, and a topiary with trees clipped by Portuguese gardeners, who are notably adept at the art, into all sorts of shapes, from graceful swans to a fat, pompous moose. Arum lilies, angel's trumpets, agapanthus, and tender plants found elsewhere on the island grow in abundance. However, it is the choice selections that indicate Blandy's eye for a good plant and good contacts.

Have garden; will travel

A keen plantswoman, Blandy understood the potential of this blessed climate, where plants from warm and cool temperate zones flourish alongside one another. Oak, beech, and magnolias from earlier plantings can be found alongside proteas, including *Protea cynaroides*, the king protea. Blandy regularly sailed back to South Africa, returning with plants. The gardens are now famous for their collections of plants from warm temperate habitats including South Africa, South America, Australasia, and the Macaronesian islands.

There are fine representatives of Australasian plants, such as *Telopea speciosissima*, the red-flowered Waratah; *Hymenosporum flavum*, the fragrant, yellow-flowered Australian native frangipani; and a fine *Metrosideros excelsa*, the New Zealand Christmas tree. Many plants were grown from seeds sent by gardeners with whom Blandy corresponded. She had an understandable empathy with South African species. *Leucadendron argenteum* (the "silver tree"), a rare and vulnerable species, first set seed in 1951 and over 100 new plants were raised. "I have frequently been told that the silver tree will not grow outside the Cape peninsula of South Africa, but here is living proof of their health and happiness on an Atlantic island."

Lily of the valley tree
Clethra arborea

This Madeiran native large shrub or medium-sized tree has beautiful, fragrant lily-of-the-valleylike flowers. It should be grown indoors in a conservatory or glasshouse in colder climates.

MILDRED BLANDY

Mildred Blandy created a beautiful garden owing to the benign climate and her knowledge and passion for plants. However, plants also need to be well tended; her constant attention and diligent gardeners proved that creating a great garden is not a part-time job.

→ "The Portuguese gardener is the gardener 'par excellence,' as is evidenced by his success in California, South Africa, and elsewhere. These humble people take a great pride in their work," Blandy wrote.

→ Because the soil is acidic, "large quantities of lime have to be brought over from the nearby island of Porto Santo for those crops requiring it, and the soil constantly fed with manure compost to enrich it further." Soil for vegetables constantly needs enriching and maintaining at the correct pH to ensure successful cropping. Remember when adding compost that brassicas need firm soil and to check the pH and only add lime if it is required.

→ Blandy allowed all of the camellia flowers that fall to remain around the plants; "this rich accumulation… protects the roots, retains moisture and helps with the decomposition of organic matter." However, as the flowers can carry fungal diseases, you could instead mulch round the base of plants using well-rotted organic matter or forest bark for a similar effect.

→ "So far I have mentioned the successes…but there are many failures. Gentians, lily of the valley, peonies, and Russell lupins, do not take kindly to the garden." Blandy was always willing to experiment and to accept graciously that there were plants that would just not grow as conditions were unsuitable. This applies to any garden.

Mocano
Pittosporum coriaceum

Now a critically endangered species endemic to Madeira, this small tree with leathery leaves at the branch tips and creamy, fragrant flowers is only found on northern slopes of the island.

Camellia
Camellia japonica 'Grandiflora Alba'

Her imagination was so fired by the description of the Mount Mlanje cedars (*Widdringtonia whytei*) of Nyasaland in a book by Laurens van der Post "that I obtained seed from the Forestry Department there; it has germinated freely and is shortly to be planted…" It is now critically endangered in its native habitat. The Sunken Garden was home to many other South African native plants: babianas, ixias, and sparaxis of every color, *Tritonia crocata*, *Haemanthus katherinae*, *Gladiolus tristis*, *Arctotis*, and *Aristea thyrsiflora*, "one of the loveliest of the true blue wild flowers from the Cape…seeding itself freely and liking steep drainage, it grows vigorously in an open aspect," she wrote in one of several correspondences with the *Journal of the Royal Horticultural Society* in "Notes from Fellows." There is also a reference to a hybrid *Aeonium* given to her by Major H.C.H. Pickering from seed collected in the Canary Islands.

Blandy was also aware of the importance of conservation by cultivation, describing *Isoplexis sceptrum*, the yellow foxglove of Madeira, which "has become exceedingly rare," and the excitement of seeing 20 healthy young plants in the garden of the trout hatchery in the north of the island. It is no surprise that rarities were conserved in her own garden, too: *Azorina vidalii* (Azores bellflower), *Lotus berthelotii*, which in 1884 was classed as exceedingly rare and endangered, and four specimens of *Camellia granthamiana* found in 1955 in Hong Kong, which arrived via the nurseryman "Mr. Trehane," all found a secure home in her garden.

QUINTESSENTIAL CAMELLIAS

The original *Camellia japonica* cultivars introduced from mainland Portugal and Belgium at the end of the 19th century seeded so freely at Quinta do Palheiro that there were once over 10,000 in the collection (now there are 1,000). When Blandy arrived, the entrance to the Conde's Quinta was already lined with camellias and they were planted in every available space. There were singles, doubles, reds, pinks, striped, and spotted, in endless color and size. Blandy inevitably became interested in them, adding many more of her own from the 1950s.

Gardening is particularly favorable for there is the even temperature free from extremes, a complete absence of frost, ample water supply, and fertile soil. Certainly it is a gardener's paradise, with growth being rapid and prodigious…near the house is the present garden, made over the last seventy years.

—Mildred Blandy

Roses wreathing the handrails leading down to a topiary garden create the impression that Quinta do Palheiro in Madeira is a traditional British garden, until you discover the rare and unusual tender plants introduced by the great plantswoman and gardener Mildred Blandy. A few years before Blandy began developing the garden, Florence du Cane noted that "Palheiro has been made the trial ground of many an imported treasure" (*The Flowers and Gardens of Madeira*, 1926). Admiring the results of her work it is easy to see that Mildred Blandy was perfect for the garden and the garden was perfect for Mildred Blandy.

Roberto Burle Marx

1909–1994

South America

Striped heliconia
Heliconia indica 'Striata'

The horizontal markings on 'Striata' are variable, sometimes spreading to cover almost the whole of the leaf.

The great Brazilian landscape architect Roberto Burle Marx considered himself primarily a painter, but his diverse talents extended to being a musician; a textile, jewellery, and stage-set designer; a conservationist; a plant collector and gardener. Marx had a garden from the age of seven. When he was 19, his family went to Germany to imbibe its culture and he was inspired by the Brazilian plants in the Berlin Botanic Garden. Although he studied art, an early commission led him to landscape architecture, where he used the principles of abstract art to paint landscapes and gardens, using native plants. The result was a revolutionary 20th-century Brazilian style that changed tropical gardening forever.

Burle Marx's Brazilian mother (Burle) was interested in gardening and his German-Jewish father (Marx) in design. The family was very cultured and in 1928, Roberto's father, Wilhelm Marx, took them to Germany. They rented a house in Berlin, absorbing its theater, concerts, opera, and the arts; at the same time, Roberto became a regular visitor to Berlin Botanic Garden, where he started taking an interest in the plants of his homeland. At the time Brazilian gardens were pale imitations of traditional formal European garden design, influenced by the French and Portuguese; there was no Brazilian tradition. During his visits, Burle Marx realized that the architectural shapes, vivid colors, and textures of "exotic" plants and flowers from the jungles of his homeland were an ideal palette for landscapes and gardens. With this single enlightenment, he revolutionized garden design.

On his return to Rio in 1930, he experimented with native plants. Particularly pleased with one color combination using foliage, he showed his neighbor, the architect and urban planner Lúcio Costa, and received his first commission, to design the garden for a private house. Burle Marx painted in an abstract style; now he transferred this to gardens. His ideas were refreshing and radical, a fusion of strongly architectural native vegetation and asymmetrical designs. Like other artist/gardeners, he painted his gardens. They are equally fascinating when painted or drawn on paper beforehand or in reality with the living landscape, in locations where patterns could be seen from above.

A GARDEN LABORATORY

In 1949 Burle Marx bought Sítio de Santo Antônio da Bica, near Rio, where he experimented further with his ideas. He restored the old house and chapel and went on his own expeditions into the rainforests, both in Brazil and south and central Asia, observing, collecting, and introducing new plants for his gardens that no landscaper had ever thought of using before. As a conservationist, he saw the modern garden as "the place to display, conserve, and perpetuate the existence of native species otherwise threatened in the wild." Native Brazilian plants were part of his heritage—he was appalled at the

destruction of the rainforest and was one of the first to speak out against it.

His garden was both an attractive landscape and a botanic collection, containing over 3,500 different species with over 430 philodendrons, a host of heliconias, masses of bromeliads, begonias, orchids, palms, ferns, other "exotics," and even some "temperate" species. For him, creating a garden was "a marvellous art—possibly one of the oldest manifestations of art," as well as an attempt to "regain [a] lost paradise."

With the help of his head gardener, his plants were precisely placed under his keen-eyed supervision. "Rhythm is not repetition but a matter of how one form relates to another, or how one place, texture, surface or color relates to another," he wrote in his essay "The Garden as Art Form." "The garden will be a constantly changing entity, but if it contains its own rationale, if all its parts are interrelated, then there will always be harmony." He also understood that the creation of a garden was a long and slow process, with the intervention of the gardener an essential part of its development.

Christmas heliconia
Heliconia angusta

Although vulnerable in its native habitat of southeastern Brazil, this species is widely propagated and grown in gardens, as it flowers in the festive season.

Many landscape architects have produced great works in the past few decades, but Roberto Burle Marx is an innovator, exhibiting a rare combination of artist and plantsman, having an architectural understanding, yet never being so dominated by structures that he forgets his plants in their endless variety.

—Sima Eliovson, author of *The Gardens of Roberto Burle Marx*

Bold blocks of colored plants, planted in harmony, were used to create moods; big leaves formed architectural features, emphasizing the structure of the garden, and foliage was more prominent than flowers. All were grouped together in beds whose edges looped and curved in organic shapes, creating a sense of movement. Rectangles reflected the surrounding buildings, and with changing vistas at every turn, visitors were directed through the plants. Falling water created sound and movement, and trees and shrubs were planted in natural-looking groups, the choicest as single specimen plants. In Burle Marx's garden, water plants float on tranquil pools; rounded river-washed stones contrast with slender vertical leaves and harmonize with other foliage, while walls provide a backdrop for epiphytes and a pergola supports *Strongylodon macrobotrys* (the jade vine). The whole house and garden (and where possible, the surrounding landscape) is in perfect harmony.

His legacy

In 1965, the American Institute of Architects (AIA) awarded Burle Marx its fine-arts prize, recognizing him as "the real creator of the modern garde;" in 1982, the RHS awarded him a Gold Veitch Memorial Medal. He designed and created around 3,000 landscape architectural projects worldwide and promoted the use of Brazilian native plants in landscapes, while campaigning for rainforest conservation. Fifty plants were named for him, including some he collected, such as *Heliconia burle-marxii*. Burle Marx had created not just a Brazilian style, but one for the 20th century; his ideas are inextricably linked with the spirit of modern Brazil, its landscapes, and identity. He was a warm, kind, and generous man, tolerant of faults in others; he loved life, plants, and people and was also a good cook, known for his legendary and convivial banquets. Above all, he was one of the great gardeners of our time.

Wild/red pineapple
Ananas bracteatus

A native of Argentina, Paraguay, and Brazil, it is widely cultivated as a garden plant. Small red-purple flowers emerge between the bright red bracts.

ROBERTO BURLE MARX

Roberto Burle Marx was a painter and landscaper but, of all the art forms, gardening is the most inclusive, as everyone can express their creativity. You may not be able to hold a paintbrush, but you can choose a plant and put it in the garden, in a place you find pleasing.

→ Where possible, draw a plan of your garden or use a hose or canes and string to mark out the shapes of the borders on the ground. It is easier to make changes at this point.

→ There is a saying, "There are no straight lines in nature." Curves of various sizes are naturally appealing; they soften the landscape, create a sense of movement, and encourage visitors to continue walking to see what is around the corner. They also reduce the speed at which you walk around the garden.

→ Burle Marx planted in large bold blocks, using colored foliage as a feature. If conditions don't allow the use of lush tropical plants or bromeliads, create a foliage border using colored or variegated plants, such as broadleaved evergreens or grasses, to create a similar effect.

→ Make use of water in the garden. The sound of waterfalls or fountains is soothing and contemplative; still pools reflect the sky and surroundings, so plant close to the edge of the pool to make use of the mirror effect, or plant with waterlilies.

→ Burle Marx used recycled building materials as a feature in his own garden. Recycled materials need not be decrepit; search secondhand shops or architectural salvage and recycle with style. It is not what you have, it is the way that you use it.

→ Burle Marx also created patterns using different materials (often colored) to create interest in hard landscaping. Achieve this yourself using a combination of natural stone, cobbles, setts, or colored gravel.

Ubim de espinho
Bactris bifida

This Brazilian native species is attractive, ornamental, and usually only reaches 4–5 feet. It produces sweet, tangy fruits that are gathered and eaten straight from the plant.

Plants from the humid tropics were the ideal material by which Roberto Burle Marx could turn gardens and landscapes into abstract art, their bold, architectural shapes and forms complementing the patterns he chose for his hard landscaping. From this image it is easy to see how he used this to create a distinctive style of his own. His garden in Brazil, now known as Sítio Roberto Burle Marx, which he originally bought to store his plants, became Marx's own experimental garden and is now home to over 3,500 different plant species.

Princess Lelia Caetani

1913–1977

Italy

Italian cypress
Cupressus sempervirens

This elegant conifer, a classic of Mediterranean gardens, is so widely planted that its wild origins are unknown.

The garden planted among the ruins of the ancient town of Ninfa, to the south of Rome, is the work of several generations of the Caetani family, notably Princess Lelia, a gentlewoman who turned her artistic skill to gardening. She added the final touches to what was already the beginning of an inspirational garden, taking advantage of a unique setting to create the garden as we see it today. Elegant, filled with fragrance, and festooned with flowers, there is little wonder that it is described by many as the most romantic garden in the world.

In 1297, Pope Boniface VIII (Benedetto Caetani) bought the walled town as a gift for his nephew, Pietro Caetani. There were six churches within the walls and the river Nymphaeus flowed through the town, attracting many industries, notably tanning, which relied on water and on *Myrtus communis*, wild myrtle; this still grows on the surrounding hillside. The town prospered until 1382 when it was destroyed in a civil war, leaving it a desolate, romantic ruin.

When the historian Ferdinand Gregorovius visited the town in the mid-1800s, he found it half buried in the swamp and covered by a cloak of ivy. The wild flowers caught his imagination: "A fragrant sea of flowers hovers above Ninfa…over every ruined house or church, the god of spring is waving his purple banner triumphantly, they smile and nod to you out of every empty window frame, they besiege all the doors…you fling yourself down into this ocean of flowers, quite intoxicated by their fragrance." Another visitor wrote: "Only the song of mid-May's nightingales broke the silence of beautiful desolation, half veiled in the ivy for centuries."

From derelict town to glorious garden

Ada Bootle-Wilbraham married Duke Onorato Caetani in 1867. They had five boys and one girl; she took them to picnic among the ruins of Ninfa from time to time. She introduced bamboos, which still remain, and planted roses by the walls during the 1920s.

Their son Prince Gelasio, who inherited Ninfa, was a true Renaissance man. Among his many talents he was a talented sculptor, pianist, and mining and metallurgical engineer; he ran a successful business in San Francisco, lectured at Harvard University, and later became the Italian Ambassador to Washington.

He also began the restoration of Ninfa and planted the structure of the garden. In 1905 he stabilized its walls, restored the town hall (as his home) and the great tower, strengthened the riverbank, and added a bridge over the river. In 1918 he employed Austrian prisoners of war to remove debris from the town and to clear brambles and self-seeded trees whose roots damaged the walls, leaving some of the ivy to retain a sense of antiquity. Gelasio loved planting trees; pines, *Magnolia grandiflora* ("Southern Magnolia"), with its magnificent glossy evergreen leaves and large creamy flowers, *Quercus ilex* (evergreen holm oak), and *Juglans nigra* (black walnut) were planted for shelter and privacy. He added *Cupressus sempervirens*, the Italian cypress, including an avenue following the line of the old main street; as he planted them, his mother added climbing roses to clamber up them. He is also said to have introduced plants he had seen in California; not one is out of place. "Don Gelasio's planting of the large trees and particularly the cypresses is superb…lending dignity to the landscape and focal points to the design," wrote Lord Skelmersdale in the *Journal of the Royal Horticultural Society*. The structure of the ruins and the trees Gelasio planted were the foundation of the garden; by 1930, the whole town had been cleared of rubble and the grass mown, the area within the walls being about 14 acres.

The garden then passed to Gelasio's nephew, Camillo, and his American-born wife, the journalist and art collector Marguerite Chapin. Camillo diverted the lake and springs into innumerable streamlets and channels so they run throughout the garden, creating a sense of cool and providing irrigation in dry summers; she planted trees, shrubs, roses, a group of magnolias, and two avocado trees.

Luculia
Luculia gratissima

This semi-evergreen with gorgeously scented flowers from the foothills of the Himalayas needs the protection of a conservatory in cooler climates, good drainage, and acidic soil.

PRINCESS LELIA CAETANI

Anyone blessed with such a refined vision as Princess Lelia certainly has an advantage when it comes to gardening. However, it is an inclusive pursuit; anyone can create something that they appreciate and enjoy. Just go out, plant something, and love it!

→ Roses were allowed to climb through trees at Ninfa. Make sure that climbing plants are not too vigorous for the tree or it may be damaged by the weight, particularly if it is old. When planting, position the climber where the prevailing wind blows it against the tree, and plant on the edge of the crown rather than at the base of a mature tree, where there is less root competition, then train it up into the tree on a rope or cane.

→ Princess Lelia was aware of views and spaces, realizing that placing plants in relation to their surroundings affects the character of the plant. A mulberry by an old wall, for example, reinforces the sense of history, or perhaps the texture of a plant's leaf complements or contrasts with a wall.

→ A rock garden isn't only a setting for alpines. They can also feature plants from sun-baked rocky Mediterranean hillsides. Along with the cyclamen and *Sternbergia lutea* that Princess Lelia planted, you could add species tulips or mullein on an open, sunny site.

→ When planting specimen shrubs or large trees, put a cane in the planting spot beforehand and imagine the tree in place, as it would be when planted, then later in its life, bearing in mind the ultimate shape and size of the plant, the shadow it will cast, and how far the roots will spread. This is particularly important with ornamental cherries, which have roots near or on the surface and can cause problems in lawns.

→ To help your imagination, sketch the area around where the cane is placed, then draw in the shape of the mature tree, using a rough scale. Alternatively, take a photograph. You can then decide if it is "the right plant for the right place."

→ Look everywhere for inspiration. Princess Lelia walked in the mountains and observed native plants growing along the river. Anything can be a catalyst for planting or design ideas: shapes, forms, places, or plants.

Fringed iris
Iris japonica

This dainty plant with attractively patterned flowers spreads rapidly in a sheltered spot in sun, part shade, and moist soil. Remove the leaves when they die back after flowering.

LELIA, THE GARDENING PRINCESS

Princess Lelia created the romantic soul of the garden we know today. She and her husband, the Hon. Hubert Howard, committed their lives to improving the estate and garden. Even as a child, Princess Lelia gardened with her mother, who wrote of "digging and sowing and transplanting and having a wonderful time." When she took over its management in the late 1940s, her intention was to fill the garden not only with trees and shrubs but with roses, annuals, and unusual tender plants, too. As an artist, she often depicted the garden as it was and how she saw it in the future. Once, she painted a view, filling a space with two *Cercis siliquastrum* (Judas trees) and an Italian cypress that were there in her mind; she was so pleased with the result that they were later planted.

She pondered every placement, planted carefully, looked at it critically, and moved it when necessary. Despite the predominantly alkaline soil, she planted magnolias (*M. sargentiana*, *M. sieboldii*, and *M. kobus*) and camellias in pockets of humus-rich soil, and rhododendrons, mainly in raised beds. Trees and shrubs were allowed their own growing space; pruning was minimal and paths were sometimes rerouted to accommodate their spread.

Princess Lelia loved rock gardens, collecting almost all of the English books on the subject. She and her mother created one on the remains of a stone wall, planting it with shrubby dwarf pomegranates, rock roses, dianthus, and bright annual eschscholtzias, *Sternbergia lutea*, and *Cyclamen repandum*

and *C. neapolitanum*. Below was the nursery where she nurtured cuttings and young plants.

Nature was often her inspiration. On one occasion, she planted *Iris pseudacorus* (yellow flag iris) in the margin of the river because it grew in the damp meadows of the countryside beyond. Everything she did was the result of prolonged evaluation and contemplation; sites and plants were constantly analyzed and assessed to create a garden that looked natural, though everything was precisely placed.

After Princess Lelia's death in 1977, Hubert Howard set about improving his own knowledge of plants and joined the RHS, reading books and magazines and the *Journal of the Royal Horticultural Society*; he wanted plants of any kind, but in line with Princess Lelia's wishes. He increased the collection of flowering cherries with varieties such as *Prunus* 'Accolade', *P.* 'Shimizu-zakura', and *P.* 'Taihaku', adding beauty to each spring.

This unique garden, its ancient walls draped with climbers and its streets filled with flowers, is the work of generations of Caetani gardeners, but the one whose vision imbued it with the spirit of romance was the artist-gardener Princess Lelia Caetani.

In the year 1920, I Gelasio Caetani planted the trees at Ninfa and restored this room, which was threatening to fall into ruin.

—A plaque in the garden

A garden with a river running through it has its own seductive beauty; surround it with the ruins of an ancient town and you have the perfect template for a romantic garden. Ninfa needed the sensitive hand of an artist to capitalize on its unique location and realize its glory; this came in the form of Princess Lelia Caetani, who refined the work of earlier generations of the Caetani family to create the garden as it is today. It is now gardened using minimal intervention, creating a rich wildlife habitat where gardening and nature walk in harmony.

Princess Greta Sturdza

1915–2009

France

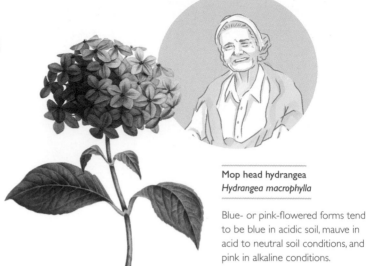

Mop head hydrangea
Hydrangea macrophylla

Blue- or pink-flowered forms tend to be blue in acidic soil, mauve in acid to neutral soil conditions, and pink in alkaline conditions.

With considerable industry, vision, and determination, plantswoman and gardener Greta Sturdza cleared an overgrown site on the French coast near Dieppe and turned it into a glorious woodland garden. She used a wide range of plants, exacting standards of horticulture, and her own ideas on their care and cultivation; the plants look healthy, the garden has flourished, and looks mature beyond its years. She has created the perfect home for many plants, among them magnolias, cherries, and rhododendrons, which show their gratitude for her care and attention with annual displays of beauty.

When the Communists came to power after the Second World War, Greta and her husband Prince George fled Romania with their family. They finally settled in Normandy, on the edge of wind-swept white cliffs on the coast near Dieppe. At first they thought the site was flat, but on clearing a jungle of scrub covered with brambles and bracken, they found to their delight that it was undulating and sloped steeply to the east.

The soil was rich acidic clay, with an inclination to waterlogging; the site was also subject to high winds and frost pockets in the valley. "I had never gardened on acidic soil before and knew very little about the plants that grew in such conditions. But I soon started to visit some gardeners in the south of England who had created fine gardens on similar soils," she wrote. She visited Britain for plants; she bought hellebores from John Massey and met Sir Harold Hillier, Collingwood "Cherry" Ingram, and Lionel Fortescue, whom she describes as "the greatest gardener and plantsman I have known;" he became a lifelong mentor and friend.

A GARDEN OF FOUR SEASONS

Le Vasterival became one of the finest gardens in Europe. It is full of paths and curved sweeping lawns, trees planted with skill and sensitivity, and island beds of beautifully pruned shrubs underplanted with bulbs and groundcover. Down by the stream, in the valley garden, herbaceous plants and ferns abound. There is always color and interest in flower, leaf, bark, stem, and berry, with fine collections of magnolias, flowering cherries, rhododendrons, roses, clematis—between 8,000 and 10,000 different plants; it is Princess Greta's garden for four seasons. Many gardeners have painted pictures with plants but few have done it so effectively with herbaceous perennials, trees, and shrubs.

TIPS AND TECHNIQUES

Very much a "hands-on" gardener and one of the best, Princess Greta developed techniques for planting and cultivation based on her own ideas and experience. Its success is evident in the health and quality of the plants.

Every shrub is pruned using her "transparent pruning" technique; they look natural and elegant, and with extra light shining through, there can be more perennials below. "My aim has always been to be able to put myself in the garden, look right, look left, south and north, and everywhere find something. I don't think we have a tree or shrub in the garden that hasn't been pruned," she said.

Her management of juvenile trees showed immense care, understanding, and attention to detail. Every second year, the roots are severed cleanly round the edge of the circle in the lawn to a depth of 8 inches, using a sharpened spade. The circle is then extended by 8 inches: the turf is stripped off with the spade, and the top layer is loosened with a rake to remove any surface roots, then covered with a good deep mulch of well-rotted manure and compost. This encourages the formation of fibrous feeding roots, which take up food as the mulch rots down. Turf is removed because the surface roots dislike competing with the grass for moisture and nutrients.

Christmas rose
Helleborus niger

This gorgeous pure white-flowered hellebore with pink tints to the blooms flowers anytime between Christmas and early spring. Mulch round them to prevent rain splashing soil onto the flowers.

Mulching is important too: the ground never freezes below and worms take the well-rotted organic matter into the soil, "digging" the borders without disturbing the roots; weeds are suppressed and the soil remains moist during drought. Agapanthus are mulched with pine needles in winter as a protective covering, because they do not rot away.

Sargent's cherry
Prunus sargentii

This small deciduous tree boasts pretty blossoms and outstanding red and orange fall color. Named for Professor Charles Sprague Sargent, director of Arnold Arboretum (see page 92).

With such nurture and care, it is easy to understand why the plants are so vigorous and healthy. With a keen eye for a good plant, either bought or exchanged, little wonder Princess Sturdza said: "My garden is full of memories. I think that's wonderful—to go round the garden and feel my friends around me."

PORTRAIT OF A GARDENER

Princess Sturdza was a caring character, with energy and strength of mind, who made a lasting impression on everyone. She was also a formidable taskmaster, setting exacting standards for herself and others. At the beginning, she set herself a stiff challenge: to clear brambles and bracken for two and a half hours a day, every day, and for five hours if she missed a day. Up at six every morning, she worked hard and would stride several miles around the garden daily, often with her short-handled, three-pronged hoe to remove footprints if anyone had stood in a border. Even in later years, she was extremely fit (she was once a Norwegian junior tennis champion and an Olympic-standard skier). On her two- to three-hour tours, photos were not allowed—you were expected to listen. She marched rapidly ahead carrying a step ladder, leaving groups trailing in her wake, and when there was a point to be made, she stopped, climbed up to her "pulpit," and spoke.

Princess Greta Sturdza was a Vice-President of the RHS, Honorary President of the International Dendrological Society, and was awarded a Gold Veitch Memorial Medal; but nothing gave her greater happiness than the beautiful garden with its astonishing plant collection that she developed for 50 years.

She created a huge garden out of a wilderness and she reigned over it for more than 50 years; her self-acquired knowledge of gardening was immense and taking part in one of her self-conducted tours was an unforgettable experience.

—Obituary, writer unknown

PRINCESS GRETA STURDZA

Princess Greta's passion and energy were driven by her love for the subject. There is no substitute for enthusiasm; it sees you through the hard times, makes successes even more enjoyable (and likely), and turns gardening into one of the greatest experiences.

→ Aim to come up with your own planting combinations and ideas. Princess Greta used *Digitalis purpurea*, the common foxglove, and the familiar *Sambucus nigra* 'Eva' (purple elder) to complement the pink bark of the Japanese maples in her collection, which the purist would plant as a specimen tree.

→ Dominique Cousin came from a Normandy fruit-growing family, but when the princess spotted his talent he became her personal tailleur (pruner). Together they refined the method of "transparent pruning" on all the trees and shrubs at Le Vasterival. "*La taille de transparence*" turns trees into natural, airy shapes. Far more plants grow beneath them, because thinning the branches lets in so much light. The principle is that, in nature, old trees die back from the crown so dead branches drop, allowing more light and rain to the roots; it opens up trees, reinforcing their natural shape, and you can also see what is planted behind.

→ Pruning ornamental trees is not only a cosmetic exercise, it also invigorates growth. It is important to do the job slowly and carefully; it is easier if you have someone to advise you which branches to remove. First stand back and look at the natural shape of the tree. Using clean, sharp tools, aim for an open center (think of a vase). Take out the dead and crossing wood and spindly side growth, then stop and look carefully to see what else should be removed— every tree or shrub should keep its natural shape. The timing of the operation is from mid-July until the end of August. Don't leave snags or stumps. Think about which plants should grow under the rejuvenated tree.

Clematis
Clematis viticella

Viticella-type clematis are excellent for training through trees and shrubs or growing in pots and are robust and disease resistant. Their small flowers appear from mid-summer to early fall.

Carl Ferris Miller
(Min Pyong-gal)

1921–2002

South Korea

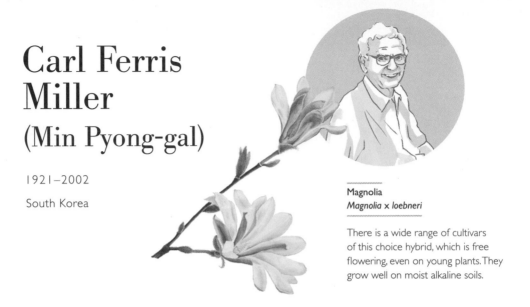

Magnolia
Magnolia x *loebneri*

There is a wide range of cultivars of this choice hybrid, which is free flowering, even on young plants. They grow well on moist alkaline soils.

Carl Ferris Miller bought land in South Korea as a favor for a friend; unsure what to do with it, he began planting trees and in doing so he discovered a passion for plants that led to the foundation of one of the world's greatest arboreta. His favorites were hollies and magnolias. He introduced many new species of plants and then donated them to other organizations, so they spread throughout the country; he also sponsored his students and gardeners, helping their education. Most of all, he was a man of great generosity, who shared his love of plants with others and to whom fame was incidental.

During the Second World War, Ferris Miller studied Japanese and served as an interpreter and translator in the US Navy. "I went home after finishing my service in 1946, but came back [to Asia] the next year. I don't know why but I fell in love with Korea from the start." He learned the language, loved its history and culture, and became a naturalized Korean in 1979 (one of only two US citizens to do so), adopting the name Min Pyong-gal.

Half his life had passed before he ignited his passion for plants, and the opportunity came quite by chance.

In 1962, a friend offered to sell Miller some land on the beach at Chollipo. Once he had bought it, he didn't know what to do with it and it remained untouched for several years; there was no property, only sand dunes, scrub pine, beach grass, and *Rosa rugosa*. He often climbed mountains, visiting remote Buddhist temples: "The monks were usually quite knowledgeable about the flora and aroused in me an interest in the vegetation."

He decided to collect trees and shrubs, rather than herbaceous plants, because "learning the names of the woody plants seemed a much more manageable job."

In 1972, he bought a traditional house on the brink of demolition, had it dismantled and reconstructed on the land, and that became his weekend home.

As the arboretum developed, he brought in horticulturists from the US and sent his own staff on internships overseas, to gardens such as Longwood in the US and RHS Garden Wisley in the UK. There were few trained Korean horticulturists at the time, so these internships created a core of trained staff for Korean parks and gardens. The arboretum is still an educational establishment for landscapers, gardeners, and plantspeople. "It is a way of repaying Korea for having me live here and having me as a Korean citizen," he once said.

HOLLY AND MAGNOLIA

Miller bought more and more land, so the site now covers 64 acres. Inevitably, many of his original selections were experimental, but there were far more successes than failures.

His great loves were hollies and magnolia. There are now over 400 taxa in the holly collection and 190 different magnolias, one of the largest collections in the world.

Euonymus
Euonymus japonicus

This glossy-leaved evergreen has contributed some excellent plants to gardens, particularly variegated forms. Thriving in sun or shade, they are especially at home in towns or coastal areas.

In its heyday, the arboretum received a thousand plants a year. Although living in Seoul, Miller travelled three hours to the arboretum every weekend, often with boxes of plants, and the first thing he did on arrival was unwrap them. Miller loved surprises: "His eyes twinkled as he carefully examined each of the plants, from nurseries in America, New Zealand, and Hillier Nurseries in England" and "He examined plant specimens, admiring them as if he was talking to them," wrote Kunso Kim, Head of Collections and Curator at the Morton Arboretum, Illinois, in a memorial tribute. Miller published an annual *Index seminum* (seed list), exchanging seed with other botanic gardens, and was a regular visitor at the annual auction of the American Holly Society—though some of his purchases, lacking the relevant certification, never made it into South Korea.

He also undertook his own annual seed-collecting trips to remote parts of South Korea in fall, boosting his collection of native species. Journeys were protracted and Miller rarely reached his destinations in daylight as he constantly stopped to look at plants. It was on one such trip that he spotted the natural hybrid *Ilex × wandoensis* (Wando holly) on Wando Island in the southwestern part of the country; an *allée* of this and *Ilex cornuta* now leads into the arboretum.

Miller also made his own selections. *Euonymus japonicus* 'Chollipo', a large dense upright shrub with broadly edged bright yellow leaves, was introduced from the arboretum to the US in 1985. It also received an AGM from the RHS in 2002, after its introduction into Britain. He selected *Magnolia × loebneri* 'Raspberry Fun' (a seedling of 'Leonard Messel') in 1987. Others were named in his honor by nurserymen, such as *Magnolia sieboldii* 'Min Pyong-gal' and a form of the rare Daimio oak, *Quercus dentata* 'Carl Ferris Miller', a vigorous rounded tree with large, thick leaves.

Chollipo Arboretum's collection of over 7,000 different species has been described

Oyama magnolia
Magnolia sieboldii

as "a plantsman's dream, one of the best labelled gardens ever visited. Both native and exotic species were there in abundance." About three quarters of the plants were new to South Korea; he donated trees to organizations rather than keeping them for himself: "I want these trees spread around Korea," he said.

Miller had a brilliant memory; he knew the name of every plant in the arboretum in Latin, Korean, and English. He was generous with his time and money, good-humored, and determined. Intrigued by its numbers and puzzles, he played bridge for the Korean national team; a friend once found him playing bridge in a sweater with a large holly leaf motif, which he also wore when plant collecting. He spent his time, energy, and fortune making Chollipo great, receiving a Gold Veitch Memorial Medal from the RHS in 1988 in recognition of his work.

The arboretum remains as a monument to one man's generosity, his love of woody plants, and the adopted country he made his own.

I had no idea I would create an arboretum recognized by international horticultural societies, or that I would give up my nationality, or that I would be awarded the highest honor the Korean government can bestow on a civilian. I just wanted to plant a few trees.

—Carl Ferris Miller

CARL FERRIS MILLER

When considering trees and shrubs for your own garden, a visit to an arboretum gives you a chance to "look before you buy" at the desirable features of each plant, including its ultimate height and spread when mature.

→ Carl Ferris Miller was particularly fond of magnolias and hollies, both popular garden plants. Don't just consider the common varieties—there is a wide range to suit all gardens.

→ Miller bought his plants from specialists such as Hillier Nurseries of Winchester. Specialist nurseries offer a wider range of plants than garden centers, especially those that are rare and unusual. They are also a source of reliable advice from experts who are passionate and knowledgeable about their plants.

→ When growing magnolias, make sure the conditions are suitable. Many are early flowering and in some years are damaged by frost. Breeding and selection has increased the range of flowering times; to avoid disappointment, make sure you research carefully and buy one that flowers after the last frost in your area.

→ Many hollies need male and female plants to produce berries. If you want berries from one plant, grow self-pollinating selections such as *Ilex aquifolium* 'JC van Tol'.

→ Miller selected his own varieties from seed-raised plants or "sports" that appeared in the arboretum. With a large collection of plants to choose from, this was more likely to happen, but chance seedlings or mutations may occur in your own garden and you may find an introduction of your own. This usually necessitates growing for several years to ensure that it is a "good" plant, then negotiating favorable terms with a nursery.

Holly
Ilex aquifolium

There is a huge number of cultivars with different colored berries and variegated foliage in a range of shapes and forms and spine arrangements—it is a truly shape-shifting plant.

Christopher Lloyd

1921–2006

United Kingdom

Honey bush
Melianthus major

Lloyd once said: "To be asked what is my favorite, high on my list will be *Melianthus major.*"

Christopher Lloyd was born at Great Dixter, Northiam, Sussex. His father, Nathaniel Lloyd, employed Arts and Crafts architect Edwin Lutyens to remodel the 15th-century house and lay out the structure of the 6 acres where Christopher and his mother, Daisy, created a garden. "I learned from her the business of gardening from as far back as I can remember," he recalled on *Desert Island Discs* on BBC Radio 4. Christopher spent most of his life experimenting in this laboratory of living color, documenting his experiences at Great Dixter in books, magazines, and newspaper columns; his informed, opinionated style, based on personal experience, gained him many followers worldwide.

Daisy Lloyd, a keen plantswoman and gardener, was friends with celebrated gardener Vita Sackville-West and her husband, Harold Nicolson, of Sissinghurst Castle (see page 122). When her son Christopher was around seven, she took him to meet another friend, the great gardener and writer Gertrude Jekyll (see page 60). "[Jekyll] was on her knees splitting polyanthus with a knife after they had flowered. I must have impressed her because she blessed me as we left and said she hoped I would grow up to be a great

gardener." However, there was a hiatus of several years before his gardening career began. Christopher Lloyd went up to Cambridge University, was called up for National Service, studied decorative horticulture, and lectured at Wye College, Kent, before finally returning to Great Dixter in 1954.

SORCERER AT WORK

What set Lloyd apart was his eagerness to experiment with color. He was constantly challenging, scheming, playing; many of his ideas were radical. He was never concerned about the opinions of others; he just wanted to find a way to make all colors and textures work together. Christopher loved vibrancy in leaf and flower, relishing the opportunity to mix plants that had never been companions before in a joyous riot of desegregation, demonstrated to perfection in his long border at Great Dixter. Here there is white-flowered cow-parsleylike common bishop's weed (*Ammi majus*), elegant double mauve-flowered larkspur "Sublime Lilac," and rich blue *Salvia guaranitica* 'Blue Enigma' in front of a pale-leaved privet. *Verbascum olympicum*, with their candelabra of pale yellow flowers, became the exclamation marks among clumps of dahlias.

Unorthodox plant combinations often shocked those who saw or read about them. Yellow and pink were a favorite: "They often work very well together, you see them in the same flower and nobody says that's wrong," he opined. He wrote of a bush of mauve-pink *Daphne mezereum*, underplanted with a carpet of *Crocus × luteus*, in zinging bright orange-yellow. "Two colors may be shouting at each other," he wrote, "but they are shouting for joy." Lloyd also believed

that plants should assert their right to roam. If a yellow spike of mullein (*Verbascum* sp.) popped up in a clump of bright pink phlox, it should be welcomed. "Hurrah for vulgarity!" he wrote.

In 1993, Lloyd again shocked the Establishment by heaving out the ancient plants from his Lutyens rose garden (he later described the "wonderful tearing of roots" as his head gardener Fergus Garrett dug them out) and replacing them with an exotic border of hardy bananas, dahlias, and cannas. However, it was not all about flamboyancy and never just for effect. Alongside the bright borders were meadows and orchards, a millefleurs of bulbs and wild flowers and topiary, a legacy of his father, all carefully tended, modified or improved.

Mauve-pink *Daphne mezereum* and golden yellow *Crocus × luteus*: one of Lloyd's many color combinations.

Spurge laurel
Daphne mezereum

Golden yellow crocus
Crocus × luteus

CHRISTOPHER LLOYD

Christopher Lloyd was free thinking and unrestrained by convention. This open-minded approach meant that he was willing to take risks and experiment—the only way to expand creative horizons.

→ Lloyd increased his range of options by being bold and embracing bright colors. Contrasts can be softened by surrounding them with subdued colors or neutral shades of green.

→ Lloyd was a keen botanist and plantsman, always discovering, buying, and trying new plants. This enabled him to increase his color palette and possibilities for new ideas.

→ Never be in a garden, your own or someone else's, without a weatherproof notebook to record your observations, note plant names, and jot down reminders of things to do. Lloyd was a great believer in such a discipline, which he is said to have learned from the Japanese. He would not answer a plant-related question unless the person asking the question had a notebook with them.

→ It is not always necessary to follow "the smallest at the front, tallest at the back" rule when planting borders. Use transparent plants such as *Verbena bonariensis* at the front, or create channels of low growth running from the front to the back.

Dahlia
Dahlia pinnata

This species is found in the wild in Mexico at high altitude. The first seed was collected around 1789 and sent to the Royal Botanic Gardens, Madrid.

→ Mix plants to create a variety of texture, not just perennials but also shrubs, climbers, and annuals. "I would mix any type of plant in a border rather than go for the old herbaceous which people love to categorize, for safety."

→ Like the plants at Great Dixter, make sure yours are always well grown, vigorous, and healthy so they make an impact and are remembered for the right reasons. You do not want your plants to be poor and disease ridden.

→ "I am not keen on lawns; they take a lot of upkeep. A decent lawn is as much work as a complicated border and it seems to me unforgivable that part of gardening should be boring and labor intensive. Over time I have converted my lawns to other things. My mother loved meadow gardening, using crocus, snowdrops, daffodils— everything is allowed to seed."

A GREAT GARDEN WRITER

Lloyd attended Rugby School, where tradition dictated that pupils wrote home after every Sunday lunch. He continued to be a prolific letter writer, his correspondence with fellow gardener Beth Chatto (see page 170) being published in *Dear Friend and Gardener* (1998). Lloyd's experiments and experiences at Great Dixter were the basis of 25 books, notably *The Well-Tempered Garden* (1970) and *Foliage Plants* (1973), as well as articles in the *Guardian* and *Observer* newspapers and magazines, and a column in *Country Life*, for whom he wrote without missing a week from May 2, 1963 to 2005. "I'm called opinionated," he said, "but I think it is very boring if you don't have opinions."

In recognition of his services to horticulture, Lloyd received the RHS Victoria Medal of Honor in 1997, an OBE in 2000, an Honorary Doctorate from the Open University, and a lifetime achievement award from the Garden Writers' Guild (now the Garden Media Guild).

After his mother died in 1972, Lloyd made Great Dixter a social hub for gardening. He was an excellent cook whose house parties were legendary and although he could be curmudgeonly (owing to a natural shyness and pressure of work), Lloyd was also incredibly kind, exceptionally generous to all, and possessed of a great sense of fun. He was very good at plant charades and particularly liked playing the game with garden writer Anna Pavord (he was in love with her *Decaisnea fargesii*!). "I've done

Never take the "I shan't see it" attitude. By exercising a little vision you will come to realize that the tree, which has a possible future, perhaps a great one, may be more important than yourself, nearing your end.

—Christopher Lloyd

what I can in my lifetime. When you've been in a place eighty years, you should have improved things," he said on Desert Island Discs; his work inspired and improved the lives of many gardeners.

THE FUTURE OF GREAT DIXTER

In 1992 Christopher Lloyd hired Fergus Garrett as head gardener at Great Dixter. Working together for nearly 15 years, Lloyd and Garrett constantly analyzed the garden. During their daily rounds, they asked questions such as "Is it worth it? Does it grow well? Does it stand on its own? Is this idea working?" About half the questions were dealt with immediately, the remainder stored up for the appropriate time of year.

Since Lloyd's death, Garrett has used the past to inform the future and has introduced ideas of his own, expanding the boundaries. There are more self-seeded plants and an emphasis on "link plants," including bronze fennel and *Verbena bonariensis*, to bind schemes together. Although he is gone, Lloyd's presence still remains: "Christo is not hovering over my shoulder. He is by my side," said Garrett.

This colorful planting at the end of the Long Border in the garden at Great Dixter encapsulates the spirit and ideals of Christopher Lloyd, who eschewed the safety of pastels and took a mischievous delight in being bold. He had the great gardening skill and knowledge of plants required to successfully combine the tones of orange and yellow *Kniphofia uvaria* 'Nobilis', red hot poker, *Phlox paniculata* 'Doghouse Pink', perennial phlox (bottom left), with the yellow spikes of *Verbascum olympicum*, backed with dark purple *Clematis jackmanii* 'Superba', Jackman's clematis (top left), into a triumphant chorus of color.

Beth Chatto

b. 1923

United Kingdom

Mount Etna broom
Genista aetnensis

This large shrub or small tree produces leafless shoots and masses of fragrant flowers in summer. It thrives in hot, dry Mediterranean conditions.

In 1967, having trained as a teacher and raised her family, 40-year-old Beth Chatto, a keen amateur gardener, decided to officially open a nursery, growing little-known plants. A house move had already taken her to a site on a range of poor soils in one of the driest parts of Britain, and it was here that she made her name, transforming rough scrubland that most traditional gardeners would deem "impossible" into an elegantly artistic garden, based on a lifetime's meticulous research by her husband, Andrew, on plant ecology and communities.

In 1960, the Chattos' new house was built on the least fertile corner of what was once part of their fruit farm, an overgrown wasteland of willow, bramble, and blackthorn. The site near Colchester was a long spring-fed hollow with waterlogged soil, surrounded by a sun-baked mix of sand and gravel, with an average rainfall of around 20 inches per year. Most gardeners would be cowed by such a challenge, "But it was the extreme variation in growing conditions which intrigued us, the possibility lying before us of growing…plants adapted by nature to different situations."

THE SCIENCE

Their confidence and optimism was born of the work of Beth's husband, Andrew, whose remarkable foresight and dedication in researching the habitats of garden plants is the foundation of the garden. Without his oft overlooked contribution, it is unlikely that the garden would have taken the form it holds today.

Andrew Chatto was a truly remarkable man. Until the age of 90, he methodically documented the native habitats of garden plants in the temperate world, from desert to forest and Arctic. His aim was that gardeners should understand more about their plants by knowing where they grow in the wild, so they could put the "right plant in the right place" and get the best from them in cultivation. He also documented their communities, identifying plants that grew alongside each other, so plant associations could be created in borders. His research enabled gardens to be created sustainably, with minimal ongoing intervention, and provided an answer to the question of "problem areas." The Beth Chatto Gardens are also an early example of eco gardening. The garden is a symbiosis of their two talents: Mr. Chatto provided the science, Mrs. Chatto selected her plants and planted the pictures.

THE ART

Although she has no formal training in horticulture or design, Chatto was a founder member of Colchester Flower Club, a skilled demonstrator, and a friend of the late artist and plantsman Sir Cedric Morris, who applied principles of art to his garden at Benton End in Suffolk.

Her first visit to his garden was "a life-changing experience." In a large room in a barn were "dramatic paintings of birds, landscapes, flowers, and vegetables—it was as if I was seeing colors, textures, and shapes for the first time," she wrote. As for the garden, it wasn't until after his death that she realized "Cedric's garden was an extension of his palette. It was not a planned painting but a collection of colors, shapes, and textures emerging and fading with the seasons."

THE GARDEN

The soils and microclimates of the 5 acre garden Chatto created are reflected in the titles of some of her books, dispensing practical advice based on experience. *The Dry Garden* (1978), *The Damp Garden* (1982), *Beth Chatto's Gravel Garden* (2000), and *Beth Chatto's Woodland Garden* (2002) document the learning curve she experienced during the garden's development.

There are plenty of clues to the practical nature of the advice in *The Damp Garden*. She writes of paths: "It is wise to follow the contours of the land, and let your path take the gentler slopes...before long you will be pushing a heavy barrow up or down the slope." Comments about plants reveal hopes and frustrations, as with *Smilacina racemosa*, false spikenard: "Mine give every promise that later I shall see little berries, like "vitrified drops

Kara Tau garlic
Allium karataviense

This bulb produces a pair of furrowed leaves at ground level, with the spherical flowerhead sitting in the center. The white, pink-tinted flowers against the blue-gray leaves is an unforgettable combination.

of blood," but the promises drop off. Perhaps they need a moister atmosphere."

One of the most well-known areas, the Gravel Garden, was created from the car park in 1991. It is encouraging to realize that the area was prepared much as you would expect from a "hands-on gardener"— and as you would do at home. Beds were laid out with hosepipes, then modified until the individual shapes related to each other, creating a cohesive overall design. After this, a rough sketch was made and key measurements, such as the widest and narrowest points, were taken. The soil was then broken up and organic matter incorporated before planting. Even drought-tolerant plants need help to become established, particularly on free-draining impoverished soils, so they were plunged in water to soak the rootball before planting.

Throughout the garden a wide range of plants is used, from annuals to bulbs, woody perennials, climbers, woodland plants, and aquatics. *Tropaeolum speciosum*, flame creeper, at home in areas where the soil is moist, was planted in the Damp Garden; humidity was raised by the surrounding vegetation and springs kept the ground moist, a prerequisite for its survival.

On the design side, planting is naturalistic, with texture and form to the fore. The principle of the asymmetrical triangle, a basis of flower arranging, is implemented in the garden on a larger scale. Tall plants paint the sky and angled lines of shrubs pull the triangle out to a long base, with the emphasis on creating a good framework of branches and greenery, then adding the flowers. Visiting the garden is a great way to learn.

Chatto has received many awards recognizing her artistic skill and the advancement of horticulture, including an OBE, the RHS Victoria Medal of Honor, The Lawrence Medal, awarded annually for the best exhibit at a Royal Horticultural Society Show in 1987, and, from 1977, a run of ten successive Chelsea gold medals.

Beth Chatto is not only a plantsperson and practical gardener but an artist whose work gardeners can learn from and enjoy. She continues the painterly style passed down through William Robinson, Gertrude Jekyll, and through her "old friend and teacher," the painter and plantsman Sir Cedric Morris; but underpinning it all is the scientific work of Andrew Chatto.

We may have a wider approach to garden design if we have been helped to appreciate other forms of art: to be aware of basic principles—balance, repetition, harmony, and simplicity—which apply to all forms of creativity. To look for these ideas in painting and architecture, or hear them in music, has certainly influenced me as much as knowing whether to put a plant in the shade, or in full sun.
—Beth Chatto

BETH CHATTO

The Beth Chatto Gardens prove that a garden can be created anywhere if the correct plants are used. Her practical experience and advice is both educational and accessible.

→ When creating a garden in this way you have to "accept the existing conditions, choose the right plants for that location, and work with what is there—plants must fend for themselves." It is using what is there that is the key, rather than taking the option that most gardeners do, of making wholesale changes to the soil and site to suit the traditional style.

→ Plants selected in the Chatto garden come from similar conditions around the world. For example, Mediterranean habitats are represented not just in the Mediterranean but in California, parts of Western and South Australia, southwest South Africa, and central Chile, too. Suitable plants growing in any of these places can be planted in a Mediterranean habitat. (Mr. Chatto's research is available on the Beth Chatto Gardens website: www.bethchatto.co.uk.)

→ Everything in the garden is arranged so it looks good now and contributes in the future—one year, five years, ten years—so it is advantageous to know your plants.

→ "I learnt from Cedric to remove the ripe seedheads of *Nectaroscordum siculum*, Sicilian honey garlic, before it seeded. A few accidentals are desirable…but left to itself the lush leaves of this curiously beautiful allium can smother the young foliage of neighboring plants." Always consider the density and spread of the leaves when planting and ensure there is enough space so that they don't swamp less vigorous plants nearby.

→ Because the plants are chosen to suit the habitats in Beth Chatto's garden, they are never watered. You may not have planted in this way, but it is still possible to conserve water in your garden by watering

Daffodils
Narcissus sp.

"I prefer daffodils in a woodland setting, since most increase well and create a natural effect. Some hybrids retain the grace of their parents and are so attractive I could not be without them," wrote Chatto.

effectively. Water in the evening, so it soaks in overnight; water the soil, not the plants; mulch to conserve moisture; and use water-efficient drip irrigation. Save water for your plants from the roofs of garden buildings and your home.

Geoffrey Smith

1928–2009

United Kingdom

Spring gentian
Gentiana verna

This tiny perennial produces masses of ultramarine blue flowers with white throats opening in late spring/early summer.

Geoffrey Smith was born in a gardener's cottage in Swaledale, Yorkshire, surrounded by beautiful countryside. On leaving school, he worked with his father before attending the local horticultural college, where he graduated as "student of the year." Aged 26, he became superintendent at what is now the RHS garden Harlow Carr, Harrogate, before becoming a successful writer and broadcaster. His encyclopedic knowledge of plants, broad experience of gardening, and lyrical use of English endeared Geoffrey to all who knew or heard him.

Geoffrey Smith was born at Barningham Park, Swaledale, Yorkshire, where his father, Frederick, was head gardener. Even as a child, he loved being outdoors, as he recalled on BBC Radio 4's *Desert Island Discs*. "My mother once commented to the vicar, 'What child would miss his dinner to watch red squirrels play?'"

Smith won a scholarship to boarding school, then worked as a forester before returning to work with his father, receiving "an education in the fundamentals of gardening." The estate boasted terraces, a rock garden, a walled vegetable garden with fruit around the walls, grapes in the glasshouse, and rhododendrons and daffodils in the park. He planted, pruned, and propagated; he could turn his hand to anything.

Six years later, realizing the need to expand his horizons, Smith applied to what is now Askham Bryan College of Agriculture and Horticulture, near York. He was awarded Student of the Year and the Yorkshire Agricultural Society's gold medal.

YOUNG AND IN DEMAND

In 1950, aged 26, Smith was appointed superintendent at the Northern Horticultural Society Gardens, Harlow Carr, where he remained for 20 years. The garden, still in its infancy at that time, was little more than a piece of rough ground with a basic master plan, but with his enthusiasm, knowledge, and practical expertise, it was destined to succeed. Smith was very much a "hands-on" gardener; when he arrived "…he took his coat off and got down to hard graft" (from an article by Ronald Wilkinson, 1974). Shelter belts and hedges were planted, the beck running through the garden was landscaped, and the Streamside Garden created. There was no machinery, so they often improvised. Wilkinson also noted that "Geoffrey Smith and his team…built, with their own hands, four pack horse bridges from coping stones." There was a general reluctance to test their strength

but shortly after they were completed a party of about 30 ladies posed on a parapet for pictures. "I must confess I was dead worried at the time. But it held. I've never had any qualms since then about our bridges," he said. Plants such as gunneras, candelabra primulas, and irises flourished; moisture-loving species were planted by the stream, while those thriving in dry conditions were placed further up the bank.

In 1954, they began building the Sandstone Rock Garden and in 1959, the Limestone Rock Garden; both were constructed by hand, using crowbars, pulleys, and ropes. The kitchen garden, trials area, and peat terraces followed. With guiding hands from great gardeners, donations of plants from nurserymen, and support from a knowledgeable committee, a garden slowly emerged.

In 1960, the director of gardens at the Northern Horticultural Society Gardens at Harlow Carr wrote: "My general instructions are given to Mr. Geoffrey Smith…the detail work as to precisely how, when, and in what way they shall be carried out is his responsibility. That is very largely why I will say once again…that what success we have achieved in building our Gardens is due in a very large measure to him…"

Smith often said that the role of Harlow Carr was to provide a demonstration ground where gardeners could see a vast range of

Ling heather
Calluna vulgaris

Heather is a mountain and moorland plant in Europe and Asia Minor. There are a huge number of cultivated varieties, which need acidic soil in sunshine. They retain their flower color when dried.

GEOFFREY SMITH

Asked for the gardening tip of which he was most proud, Smith cited his advice to anyone moving into a new or established garden that they should spend a year studying the site before making any alterations.

→ During one recording of *Gardeners' World*, Smith, with peach juice running down his chin, turned to camera and said, "There's no way a shop-bought peach could ever taste like that, not in a million years." Peaches can be grown in fan form against a sheltered south-facing wall or in a glasshouse. If you don't have these conditions, try other fruit such as apples or pears against a wall or as trees. Even on the smallest plot there is room to grow strawberries.

→ Before planting a tree, cut back the thick main roots by about a third to a quarter; this encourages the formation of finer feeding roots, so the tree establishes rapidly.

Rhododendron
Rhododendron cinnabarinum

This evergreen shrub with long, slender branches has leaves that are reddish brown to blue-gray underneath. The flowers are usually matt cinnabar red, giving the plant its Latin name.

→ When you plant a tree, shake it gently so the soil settles among the roots and don't plant below the soil level (that is, the level at which it was planted in the nursery or growing in a pot), or it could die.

→ Don't plant trees or shrubs when the ground is frozen or waterlogged. The soil should be crumbly so it can be worked in among the roots, by hand.

→ When planting trees or shrubs, add soil in layers, firming, but not compacting, each layer around the roots with your foot.

→ When tying in plants, use plastic or natural ties, not wire, to ensure the bark is not burned in scorching sunshine. Check regularly to ensure the ties are not cutting into the bark.

plants growing under northern conditions. He was proud of defying received horticultural wisdom, saying his was "confidence born of ignorance." "I felt that 'tender' was a term invented by southerners to stop northerners enjoying the beauty of an embothrium, framed like a pillar of scarlet against a May sky, or the virginal loveliness of *Magnolia denudata* with the soft greenness of spring all around…" he wrote in the RHS magazine, *The Garden*.

BROADCASTING

In the early 1960s a BBC producer turned up at Smith's garden and invited him to appear with Percy Thrower on *Gardening Club* (a forerunner of *Gardeners' World*, which he later presented). He was working in a tree at the time. Dropping down from the branch, he accepted on the spot.

This was followed by five other series, including *Mr. Smith's Vegetable Garden*, *Mr. Smith's Fruit Garden* and *Geoffrey Smith's World of Flowers*. Many of his books became bestsellers. Everything was written by hand then tapped out on an old manual typewriter by Smith's wife.

From 1983 Smith became a mainstay of BBC Radio 4's *Gardeners' Question Time* for two decades. He was a mine of information, his poetic turn of phrase making him one of gardening's great broadcasters. "If I am depressed, or I think the world's a filthy

Put the brown end in the soil, the green end above it, and you're in with a much better chance.

—Geoffrey Smith

place, I just go and look at a flower;" "Some people go to the whisky bottle, I go into the garden." This, combined with his "love of gardening, warmth, erudition, and wit," ensured his longevity. Ultimately, he believed that gardening could put the world to rights.

Smith was awarded Associate of Honor of the RHS, received an Honorary Masters degree from the Open University and, in 2006, the Garden Writers' Guild Lifetime Achievement Award.

He loved fell walking (thinking nothing of walking 10-12 miles per day, even in his eighties), photographing plants, and keeping bantams, although he was frustrated by their habit of having dust baths next to his rare plants. As a proud Yorkshireman, he loved cricket and was a good fast bowler, too.

"I don't need paradise," he once said. "The Yorkshire Dales will do for me."

Bird's eye primrose
Primula farinosa

The distribution of this plant extends to Europe and Asia but its conservation status in Britain is vulnerable. Only grow plants and seed from cultivated sources.

The main borders, with their view down to the Bath House, replanted several times over the years by Geoffrey Smith and other curators and head gardeners, are one of the outstanding views of the garden at Harlow Carr. Landscaping on this scale requires planting in correspondingly large blocks of color—smaller scale plantings would be ineffective. At its peak from midsummer onwards, the border contains plants of similar vigor, so one plant does not dominate. Among the herbaceous perennials used are red-flowered *Persicaria amplexicaulis* 'Blackfield', or red bistort, yellow *Crocosmia × crocosmiiflora* 'Solfatare', a montbretia, and in the background are the seedheads of *Monarda* 'Gardenview Scarlet', a bergamot.

Penelope Hobhouse

b. 1929

United Kingdom

Golden peony
Paeonia mlokosewitschii

With its cool lemon-yellow flowers, central boss of golden anthers, pale blue-gray foliage, and orange-brown fall color, this beautiful peony is always in demand.

Penelope Hobhouse is notable for her sheer breadth of interest and expertise in gardening and design. She is a gardener, garden historian, an authority on Islamic and Mughal gardens, a tour leader and lecturer, and famous for her garden restorations at Hadspen and Tintinhull. After the publication of *Color in Your Garden*, one of many books, she became a sought-after garden designer, making gardens for the late Queen Elizabeth The Queen Mother at Walmer Castle, Kent, and for the RHS at Wisley. She has a special affinity with the United States (she is part American; her grandmother came from Chicago); her work there includes the Bass Garden in Maine and an herb garden at New York Botanical Garden.

Penelope Hobhouse did not become interested in gardening until the age of 30, when she took some visitors to see the garden at Tintinhull House, across the road from where she lived. Having seen the structure and color schemes, she suddenly realized that gardening was not just about cutting nettles round the house. "That was the first moment I realized gardening was about beauty and art," she recalled on BBC Radio 4's *Desert Island Discs* in 1994. She hastily wrote to her friend John Raven, a Classics don at Cambridge and amateur botanist, with a list of yellow and white plants she wanted to grow. It was returned marked with the plants he thought would flourish in her garden. Hobhouse read widely: Christopher Lloyd, Gertrude Jekyll, William Robinson, and Californian landscape architect Thomas Church; obsessed with botanical Latin, she visited gardens, annotating her *Hillier Manual of Trees & Shrubs*, which became battered and had to be rebound.

FURTHER EDUCATION

Hobhouse became a very practical gardener. "Physical work is part of the pleasure, that is what makes a garden look beautiful. It is important to be artistic and practical: I like working outside, love weeding, and there is still a thrill from germinating seeds in the greenhouse."

Her next move was to Hadspen, to restore the family garden to its Edwardian glory. It was full of bindweed, so she worked outwards from the house, weeding, mulching, and planting trees and shrubs.

She also bought a house in Italy, educating herself on garden history and design. "I spent the next five years going to Italian gardens learning about terraces, views, gradients, structure, and space, foliage green and gray stone, it was a revelation. I had visited flowery gardens, which were too flowery for me; it was such a relief to find that gardens didn't need flowers," she said in an interview for *Gardens Illustrated*. "Italian gardens made me a better designer; I now have an historical reference point."

Another source of inspiration was visiting art galleries to view landscape paintings. "Whether a great landscape or smaller space you need to look at pictures and learn how they work, because that is what you are creating," she explained. "Unlike paintings, gardens are dynamic and need controlling. What you are looking at now won't look the same next year, and by the following year will need changing; you have to accommodate growth and the seasons, it is probably the most difficult fine art to master," she said. Her trademark is still the skilful use of textures and color within a framework that opens up to vistas and views.

Mock orange
Philadelphus 'Belle Etoile'

This deservedly popular shrub with beautiful white flowers with a purple-maroon central blotch is free flowering and deliciously fragrant, as its common name suggests.

I love gardening and if anyone says what is my recreation and pleasure, that is gardening, the other is work.

—Penelope Hobhouse

TINTINHULL

Tintinhull, her first inspiration, became her next garden. She wrote asking the National Trust if there was anything garden-related she could do to help; they offered the tenancy of Tintinhull. The 1 acre garden, formed of several "rooms," was largely the inspiration of Mrs. Phyllis Reiss, who moved there in 1933. "Her special gift," wrote garden designer Lanning Roper, "was for selecting and placing plants to create effect…if a plant is distinguished in form and texture it is used boldly and often repeated, making a unity of design."

"Mrs. Reiss didn't have a lot of money so propagated the plants herself, repeating plant associations in different rooms, proving that repetitions were better than endless new plants," Hobhouse recounted. The whole garden needed replanting,

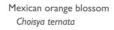

Mexican orange blossom
Choisya ternata

The cultivar 'Sundance', with bright yellow young foliage, is long flowering, the fragrant flowers appearing in spring, late summer, and fall.

so Hobhouse and her husband, Professor John Malins, continued to experiment and rejuvenate. "I thought she would have liked what was being done, she didn't have wonderful salvias or euphorbias in her day but I am sure she would have loved and planted them, because of what she had done." After 14 years, when visitor numbers reached 20,000 a year, all coming through the house, it was time to move on.

GARDEN AFTER GARDEN

The Coach House at Bettiscombe was a small walled garden on two levels. "Plants grow better because you have put them in a good location, and they behave differently to the way they would in the wild; it's called third nature," Hobhouse remarked. The front garden was green, looking out on the Dorset hills. "I am a little bit bored with the nuances of flower colors. I've done that, what interests me now are textures, form, and structure in greens and greys, like broad-leaved evergreens and *Olearias* from New Zealand, *Osmanthus, Choisya ternata*."

Her current Somerset garden, Dairy Barn, demonstrates the importance of good foliage and intriguing structure, with her hallmark—a strong formal framework, exuberant planting, with a carefully created equilibrium—still very much evident.

"At Hadspen I was recreating an Edwardian garden for the family; at Tintinhull I became a workaholic [...] Bettiscombe was for me, doing just what I wanted. My husband had died, so I fell in love with the house, the view, and the garden."

PENELOPE HOBHOUSE

Hobhouse's great knowledge of art, plants, and garden history has always informed her ideas and designs. Ideas can come from the most unexpected places, so always carry a notebook, your phone for taking photos, and keep your mind open to inspiration from any source.

→ "It is important to repeat groups of plants within a border, otherwise it just becomes a jumble of things you like; it needn't be symmetrical but it gives structure to your planting."

→ "Gardening has a physical and spiritual side, both need to be satisfied. There is a danger that too much time is spent working. Islamic gardens are for sitting and contemplation, not walking."

→ "I would say to anybody who wants to be a garden designer or writer, they must go to Italy, the Italian experience is very important."

→ "Gardening makes you an optimist, you must think ahead. Often it is the timescale that is the therapeutic element, not just the everyday practicalities."

→ "I like to plant in a very naturalistic way, then fill the borders with plants. In garden rooms the essential lines are right angles; in a naturalistic garden you follow the contours."

→ "It takes time for hedges to grow tall enough to make a garden room. You should consider 'instant hedges' up to 6' or more tall; it is costly but is worth it."

→ "Ordinary plants are very important. I love *Alchemilla mollis*, 'Catmint', and roses." It is not the plants but how they are used that counts.

→ Flower color is fleeting, and less important than the color, texture, and shapes of leaves. Foliage supplies structure for at least six months of the year. It also has greater subtlety than flowers, with less strident hues. Tintinhull was planted to Mrs. Reiss's photographs, with adaptations. "She had bright red, yellow, and orange with blue delphiniums, but I cut the heads from the delphiniums, because the color didn't work." You can do this with any plant, removing the flowers to enjoy the foliage.

Jasmine box
Phillyrea latifolia

This elegant evergreen olivelike shrub or tree is a native of the Mediterranean and Western Asia. It makes an interesting subject for cloud pruning.

The design for the Country Garden by Penelope Hobhouse at the Royal Horticultural Society garden at Wisley, Surrey, was accepted sometime during the course of 1997. The site was challenging due to the slope, so there are steps and ramps within the garden. Work began by the end of the year and was finished in 2000. She described the garden, with its wisteria arch and crabapple avenues, as "basically Robinsonian," with *Zantedeschia aethopica* 'Crowborough', the calla lily, adding a touch of simple yet sophisticated elegance.

Bev McConnell

b. 1931

New Zealand

Sweet gum
Liquidambar styraciflua

This tree has many cultivars that are valued for their stately form and fall color; some are purple, crimson, and orange—all at the same time.

Bev McConnell created Ayrlies, one of the world's great gardens, on what was once farmland near Auckland, New Zealand. Using her art school training and practical knowledge, she has composed a sequence of beautiful pictures with plants, taking advantage of the location, topography, and a climate that allows an impressive range of plants to be grown. McConnell's most recent project, the development of wetlands, has created not only an appealing landscape linking her garden to the sea but also a valuable wildlife habitat, particularly for birds.

McConnell explained: "You have to think like a painter, you need to walk a garden at all times of day and season and watch the light, the way it hits a leaf with a particular texture, or filters through the leaves; you adjust and move plants to enhance that experience. You've got to notice everything that is happening to every plant, every day." (*Wall Street Journal*, 2013.)

McConnell's parents, both gardeners, made sacrifices to send her to art school and her father was a great influence. "Every morning before breakfast on the veranda he would take me to smell the best rose in the garden, he loved his roses," she said. Her father was a GP serving a predominantly Maori population, so she grew up with an understanding and appreciation of the natural world, which is part of their culture.

In 1964, McConnell and her husband, Malcolm, moved into a 120 acre dairy farm on an unspoilt stretch of coastline, without a tree in sight. Although it remained a working farm, 3 acres were fenced off for a garden. McConnell had already compiled a list of over 500 native and exotic trees, intending to fulfil her dream. They were planted around the boundary on the first weekend of

September 1964 and the early years were dedicated to keeping them alive on wet, heavy clay. Of 64 cherries that lined the drive, few reached maturity. New Zealand natives invariably flourished, with "exotics" such as evergreen *Cryptomeria japonica* (Japanese cedar), alongside three dominant deciduous trees, North American *Liquidambar styraciflua* (sweet gum), *Taxodium distichum* (bald cypress), and *Quercus palustris* (pin oak).

MAKING THE FRAME

The garden evolved out of a desire to create a large informal country garden, where the topography dictated the shape. The plan was to create space for groups of large trees, still ponds and tumbling water, creating an air of tranquillity; it should also be filled with moments of drama, lots of heady perfume, and seasonal interest, following Vita Sackville-West's advice to "have something looking at its best every week of the year."

On a visit to England in the mid-1970s, McConnell met landscaper Oliver Briers, who moved to New Zealand with his family to help McConnell realize her ambition. He became interested in the influence of water in the garden, devising a plan to dam the valleys and create a network of streams. He built the framework of bridges and cascades, raised walls, and dug flower beds, while McConnell focused on the plants that filled the frames. She used plants to create different moods and designs, a range of sensory experiences from cool and soothing to invigorating and vibrant. The plantings are balanced and composed and the hard landscaping perfectly constructed; everything in the garden is in harmony and the garden is harmonious with its surroundings.

Kowhai
Sophora microphylla

This small tree has dark green foliage, ideal for showing off its bright yellow flowers. A hardy hybrid, 'Sun King', is often grown in gardens.

BEV McCONNELL

McConnell gardens in a challenging climate, which she has modified to her advantage by planting shelter belts to create a habitat where an astonishing range of plants will flourish. She has realized her dream by never shirking from a challenge.

→ Like all great gardeners, Bev McConnell made and maintained. The hands-on approach means you can be part of the success of your garden, direct helpers, and know what is needed. Experience and knowledge come with time and education: McConnell has gardened at Ayrlies for 50 years, so is aware of what is needed to ensure the garden's continued development and success.

→ McConnell operates under the dictum of Vita Sackville-West: "If you don't like a plant, get rid of it, you won't like it any better next year." Sometimes you have to make decisions for the good of the garden.

→ "I love color but you have to be careful; I go out at dusk so I am not distracted, then see shape and form which is more important than fleeting flowers," McConnell says.

→ McConnell painstakingly evaluates plants and removes nonperformers so only the best are on show. This approach is necessary to maintain the quality of a garden.

→ McConnell's father was an early influence. He always maintained that an element of surprise is essential in a garden, it must entice you to look round corners; you don't want to see it all at once.

→ The trees that were planted around the garden needed watering during dry weather. Any new plantings need ongoing care—watering, mulching, checking the tension of tree ties—until they are well established. Planting trees when they are small means they establish quickly and acclimatize more rapidly. They also need protection from animals. Several "wattles" were lost due to hares stripping the bark; the same problem can be caused by deer and voles.

Tree aloe
Aloe barberae

In time this striking succulent plant becomes a piece of vegetative sculpture and a beautiful focal point for a garden, with its form and attractive gray stems.

PLANTER'S PARADISE

The climate is a hybrid of maritime, warm temperate, and subtropical, where frosts are rare, humidity can be high, and there are more than 50 inches of rain every year. "No wonder it seemed obvious to lean towards bog and water gardening." (McConnell, *Ayrlies: My Story, My Garden*.) Tree growth is fast and older specimens, particularly conifers, can lose their shape.

The climate allows for an extraordinary palette of plants, from roses, magnolias, and temperate trees to aloes, tropical climbers, tree ferns, and taro. There are old roses, hybrid musks such as clear-pink, single 'Wind Chimes', and gallicas, including carmine-petalled, almost thornless 'Hippolyte' and 'Roseraie de l'Hay' with purple-crimson flowers, yellow fall color, and red fleshy fruits.

THE OUTDOOR CLASSROOM

In the early 1980s, McConnell visited some of Britain's great gardens, and could be found pacing out perennial beds to increase her knowledge of composition and how to make a perfectly balanced border. "A garden is continually changing, there is a moment of perfection but it is short lived; you are always refurbishing a garden." She learned by reading, became friends with Christopher Lloyd (see page 164), who refined her perception of color and texture, and admired

Taro
Colocasia esculenta

This plant is a "must have" for its tropical effect. Start it into growth under glass, in cooler climates, then plant out once the danger of frost has passed.

Beth Chatto's garden: "It wasn't until I saw Chatto's garden that I realized I had so much to learn. She puts plants together better than anyone I know." McConnell is a self-proclaimed romantic and tends to let things grow wild but, as Dan Hinkley (see page 214) points out, "you don't make a garden this complex and at the same time relaxed, without discipline" (*Wall Street Journal*, 2013).

At the turn of the millennium, she embarked on a major project to link the complexity of her garden to the sea, sculpting 35 acres of marsh bordered by 1,500 native plants to provide a habitat for wildlife, particularly birds.

McConnell has received the Veitch Memorial Medal, the first New Zealand woman to do so, and her garden has been designated a Garden of International Significance by the New Zealand Gardens Trust.

Gardening is an art form...the wonder of Ayrlies is that it contains a diversity of plants to rival some botanic gardens... yet is an exquisitely crafted composition that captivates all who visit.

—Jack Hobbs

Bev McConnell is aware of the "green and water" concept first used by the Moors in medieval times then adopted by the French, Italians, and English. The idea is that visitors should enter a garden through a cool green area, so that their cares of the world fall away. The cascades near the entrance at Ayrlies are flanked by grasses, shrubs, and ferns, whose arching foliage reflect the tumbling water, their soft colors enhancing the feeling of tranquillity. The eye dilates to see blue, contracts to see vivid reds, but rests when looking at green. The sound of running water adds to the sense of serenity, too.

Geoff Hamilton

1936–1996

United Kingdom

Common apple
Malus domestica

The apple is one of the most widely grown fruits in cool temperate climates. Varieties were bred to suit local and regional growing conditions.

Geoff Hamilton's family lived in Broxbourne, Hertfordshire, an area immersed in horticulture, so it is no surprise that as a 14-year-old he began working on nurseries to supplement his pocket money—and loved it! His mother wanted him to be an accountant, his father wanted him to decide for himself; so after three years of study and gaining the National Diploma in Horticulture with distinction, Hamilton set up a landscaping business. He later bought and refurbished a garden center in Northamptonshire, working so hard he wore off his fingerprints, before turning his skills to writing and broadcasting.

While the garden center was becoming established, Hamilton needed to add to his income, so he tried garden writing; he sent examples to *Garden News*, where the editor was very impressed and he received his first commissions. Hamilton lacked the ruthless edge of a businessman, and several years later his garden center went bust. He resorted to working for a commercial vegetable grower in Lincolnshire, doing unskilled manual work and enduring a period of hardship until he became editor of *Practical Gardening* magazine.

As "hands-on" horticulture was his passion, Hamilton was eager to share what he'd learned. His aim was to encourage more people into gardening and to help enthusiasts become more knowledgeable and reap the rewards. Around the same time, he rented a cottage on the Barnsdale estate in Rutland. Its 2 acres of land were ideal for "step-by-step" pictures for the magazine; from taking root cuttings to laying slabs, Hamilton demonstrated every process himself, gradually constructing a garden.

The people's gardener

Shortly after buying the garden center, Hamilton had gone to a screen test for a regional TV program, *Gardening Diary*. The other three candidates were filmed for 20 minutes, he only for five; he left thinking they weren't impressed, but they knew immediately that he was their man and he was offered the job.

At that time, BBC *Gardeners' World* often invited guest experts to feature on special projects and as editor of *Practical Gardening*, Hamilton was in a prime position. A natural in front of the camera, he instantly impressed and joined the team full time in 1979, his garden becoming the main location—it had plenty of available space. Often they decided in the morning what was to be filmed in the afternoon, so he was also expected to collect props and prepare. Hamilton bought "new" Barnsdale, a Victorian farmhouse with 5 acres of land, in 1983. There was no master plan; the garden evolved not from television projects that were fashionable, but from those he thought would help viewers by covering the subjects gardeners wanted to see.

He also presented several series of his own. *The Ornamental Kitchen Garden* (1990), *Cottage Gardens* (1995), and *Paradise Gardens* (1997) led to him building nearly 40 demonstration gardens at Barnsdale on assorted themes, including a modern estate garden, an artisan's cottage garden, a wildlife garden, and an Elizabethan vegetable garden. With a small team, he turfed, built walls, prepared borders, sowed seeds, chose plants, and created. Many were built without plans and their stories were told in more than 20 best-selling gardening books. Visitors arrived in their thousands when Barnsdale opened to the public in 1997 and many still make the pilgrimage today, to see the gardens they remember from the television and to touch the soul of a man who was their friend.

What set him apart?

Hamilton was a brilliant gardener, a natural teacher, and there was nothing he couldn't do (even his jokes were practical); viewers related to him as the man next door, showing them the "tricks of the trade" in a relaxed, friendly

Pointed leaved penstemon
Penstemon acuminatus

This leathery leaved perennial with beautiful blue, purple, or pink flowers is used in habitat restoration, on roadsides and native gardens in parts of the U.S..

manner, dressed like many a gardener in work boots and jeans. He followed a great tradition of "make do and mend" (thriftiness is part of gardening), but he was also a man of vision who realized that we could not survive as a "disposable" society; he was occupied by the idea of sustainability long before it was a global concern.

In 1996 he built a reclaimed garden. The pergola was of reclaimed oak, the raised beds and seat of railway ties; steps and floorboards were used for fencing. There was reclaimed wrought-iron fencing, and a copper immersion heater was transformed into a rose fountain; Hamilton wanted to show the public that "recycled" could be attractive and effective, too. Recycling became his theme: Chinese takeout trays became seed pots, kitchen-roll tubes were used for sowing sweet peas, orange boxes became cold frames, and scraps of underlay from the local carpet shop were transformed into barriers against cabbage root fly.

From 1986, Hamilton trialled and adopted the principles of organic gardening, then brought them to the people in a

Seedsmen reckon that their stock in trade is not seeds at all...it's optimism.

—Geoff Hamilton

Noisette rose
Rosa × noisetteana

These had their beginnings when seedlings of hybrids created by John Champney, a rice grower in Charleston, South Carolina, were passed to his friend, Philip Noisette.

matter-of-fact way, so that these ideas didn't appear odd, sensationalist, or "left field." His garden flourished under this radical new regime. "I am totally convinced this is the way forward and Barnsdale is a testament to it," he said.

Hamilton considered himself fortunate to have a job he enjoyed and his legendary sense of humor was interwoven through his everyday life. It was part of what made him so watchable; BBC staff loved working with him because it was always fun and the relaxed atmosphere at Barnsdale shone in the quality of the programs. "If you can sustain yourself doing something you want to, why shouldn't you be happy?" He was paid to do what he loved.

Hamilton died in 1996 and was buried in his local churchyard in his familiar boots and jeans (always bought from the same local agricultural suppliers—the only time he went clothes shopping). A *Cercidiphyllum japonicum*, the Japanese katsura tree, was planted by his grave. It has several seasons of interest and the leaves smell of burnt sugar in fall. Hamilton liked the fact that it was quirky, a little different, and great value for money.

GEOFF HAMILTON

Hamilton really enjoyed himself, loving every minute of his life; his favorite gardening saying was: "If it doesn't succeed, try, try again." Learning from the previous failure, then repeating the process, eventually and inevitably leads to success. Don't give up! Here are a few more of his thoughts.

→ Make life easy by working with what you have. This includes putting the "right plant in the right place," adopting organic gardening principles, recycling and cutting costs by reusing what you already have. Gardening does not have to be expensive.

→ Things don't have to be perfect, just as good as you want them to be. Pursuit of perfection creates unnecessary pressure and takes the fun out of gardening. Gardening is all about enjoyment; don't take it too seriously.

→ Create things yourself for optimum satisfaction; it is not just about the plants, it is the other things you can do as well, such as building walls or raised beds or propagating plants for free. How much satisfaction do you get from writing a check or signing your credit card?

→ Hamilton believed that with the correct information and understanding, *anyone* can do *anything*.

→ Hamilton took a very pragmatic approach, researching well before trialling anything new, be it a technique or new plant variety. He mulled over the idea and then prepared thoroughly. He was also of the opinion that new is not necessarily better. Many of the old techniques and varieties have not been replaced because they have been proven with practice and time.

Onion
Allium cepa

A vegetable of antiquity, the onion was cultivated by the Egyptians as food and also to use during the mummification process. Pliny recorded six varieties in Rome and it was highly regarded as an antiseptic.

Piet Oudolf

b. 1944

The Netherlands

Orpine
Sedum telephium

'Munstead Red', with purple tinted dark green leaves and flat flowerheads of chocolate red, was selected by Gertrude Jekyll and is used in New Perennial plantings.

Piet Oudolf trained as a garden designer, then began to look at a style that used mostly herbaceous plants and grasses in a naturalistic way. Experimentation and plant selection has expanded the number of varieties that suit this style; they are clump forming, provide maximum impact for minimal maintenance, and must look good—even when they are dead. Oudolf's innovative ideas, skill in execution, and a series of high-profile projects have established him as a leading light in the New Perennial movement.

The New Perennial movement has its roots in the work of Karl Foerster (1874–1970) in Germany and William Robinson (1838–1935) in Britain, who both believed in putting the right plant in the right place without modifying the growing conditions. They looked to nature for inspiration and aimed for balanced designs using plants that shared similar habitats in the wild. Mien Ruys, Beth Chatto, Dan Pearson, Wolfgang Oehme, and James van Sweden, who created the New American Garden style, are among those who have championed and developed this ideal. In

Edwardian Britain, late summer flowering perennials were grown for their flowers in intensively managed herbaceous borders; with the new style, many of the same plants are used for their form, not just flower, and need little maintenance, too. It is as though they have returned to their native habitat.

NEW PERENNIALS IN PRACTICE

There is a strong Dutch tradition of trimming and shaping hedges, so it is no surprise to find Piet Oudolf's 1 acre garden at Hummelo has a strong structure of beech and yew, clipped into contemporary shapes—undulating curtains, tables, and columns—as a background to the soft herbaceous nature of his planting.

The garden is divided by a central path so narrow you feel as though you are part of the garden, with three offset circular beds. Walk round each island and the view constantly changes; plants need to look good from all angles, not just from one side. "I don't draw plans for our own garden, I make lists, work out an idea in my head, and set the plants out by eye," Oudolf explains.

His garden is a place of experimentation and change, where new ideas are developed; he also learns and returns from other projects then changes his own garden, too. For example, he has always used late-flowering perennials, but since visiting the Prairies in 2001 he has used many more.

ARTIST AT WORK

Oudolf's plantings are a combination of large lush plants such as *Thalictrum* (meadow rue), *Astilbe* (false goatsbeard), *Verbascum* (mullein), and *Eryngium* (sea holly) for the structure, with more colorful, floriferous, relaxed plants such as geranium, *Stachys* (lamb's ears) and *Nepeta* (catmint) in between. (The ratio is 70 large structural plants to 30 floriferous.)

Despite being full of flowers, the borders are surprisingly low maintenance. There is no deadheading; seedheads are part of the style and there is little need for staking in spring. The main task is cutting back and

Grasses and herbaceous perennials are the foundations of New Perennial plantings. Only the best are selected; they are often close to the wild species and provide several seasons of interest.

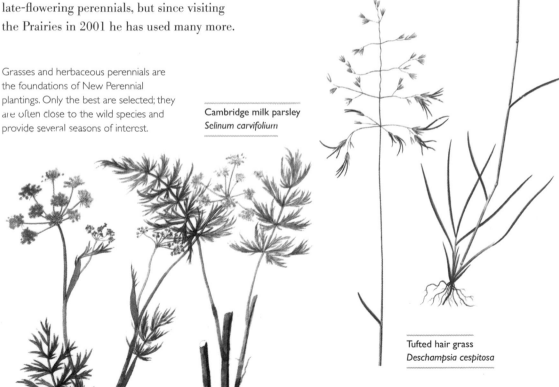

Cambridge milk parsley
Selinum carvifolium

Tufted hair grass
Deschampsia cespitosa

PIET OUDOLF

When following a particular style, like that of the New Perennial movement, it helps to visit gardens where the ideas have been put into practice, to read extensively and prepare ideas on paper before planting.

→ Flooding in the garden at Hummelo caused the death of several yews, destroying decades of growth; but herbaceous perennials recover quickly even in difficult conditions and growth is rapid if they need to be replaced. Tailor your plant choice to suit the soil, conditions, and climate.

→ The Netherlands have cold winters, so plants remain frozen in time for longer. In warmer, wetter climates, they are more likely to decay, losing their shape and form. Research thoroughly and select your plants carefully.

→ Planting schemes work because they are made up almost entirely of long-lived clump-forming perennials that don't rampage around, root aggressively, or seed, and keep their form as distinct blocks, such as heleniums, molinias, and sanguisorbas. It is one of the keys to making this work.

→ Oudolf makes regular trips to see plants in the wild, particularly in the US and Eastern Europe. Visiting their habitats would be invaluable to anyone who wants to garden this way. You don't have to travel long distances; even a local wildlife reserve or wildflower meadow offers some understanding of what plants look like in the wild.

Plume poppy
Macleaya cordata

An elegant upright plant, which, despite its size, rarely needs staking. It has lovely foliage and white flowers and does not spread, unlike some of its relatives.

→ Plants are chosen for their ecological value as food sources and habitats for insects and birds as much as for their beauty. Wildlife-friendly planting should be included in every garden.

→ Clipped yew hedging adds weight, punctuation points, and year-round sculptural interest to the design. There are traditional and contemporary shapes and styles; use one that reflects your taste and the style of your garden.

composting at the end of winter before new growth appears, while self-seeding species such as fennel or verbascum are thinned to reduce numbers and maintain the fine balance of the borders.

Almost every plant has been chosen because it is disease resistant, attracts wildlife, and is long lived, dependable, and easy to maintain. Selecting such plants is the foundation of Oudolf's work; his wife Anja runs a nursery, which grew from the need to breed, select, and experiment in order to develop his palette.

Though there are plants for earlier displays, the show reaches its crescendo in September. Most plants are at their peak and the borders are filled with flowers and foliage; structure plants and grasses dominate the scene. There are familiar and unfamiliar plants, bright bold colors blending with the fine filamentous stems of grasses. "What I like about the garden then," says Oudolf, "is that everything is at its full height and growth, there should be a sense of awe at the size of the plants."

THE LATE, LATE BORDER

The New Perennial approach requires that plants should look good when they're dead. "Dying in an interesting way is just as important as living," Oudolf says.

It came from the desire of many eco-orientated Dutch gardeners to maintain winter habitats rather than tidying borders in fall. Leaving dead leaves and stems was radical; it requires a change in perception, to see beauty in decay. The foundation of Oudolf's work is an appreciation of plant structure and that certainly is evident in the winter display.

The cornerstone is a round bed of *Miscanthus sinensis* 'Malepartus'. Grasses such as *Panicum* (switch grass), *Miscanthus* (Eulalia), *Molinia* (purple moor-grass), and *Pennisetum* (fountain grass) are not only pretty in summer, adding daintiness as they dance, but are delicate in winter, a marked contrast to the strongly defined seedheads of perennials such as *Echinacea purpurea* (coneflower) and *Digitalis ferruginea* (rusty foxglove).

Low winter sun at sunrise and sunset spotlights, or backlights, the whole scene in shades of brown, bronze, and parchment, highlighting the texture in leaves, stems, and grasses and beautifying the angles and architectural structure of dead plants. During winter's chill, hoarfrost powders the painting, emphasizing delicate details that might otherwise be missed, and spider webs embellish with their own distinctive beauty. The garden, though dead, is alive.

Oudolf has expanded the brief by tuning in to nature, gathering its glories then replanting with skill, in a New Perennial style.

I like to connect people with the processes of their own lives. What it takes humans a lifetime to experience, a plant will experience in its own yearly life cycle. In that sense, gardening is a microcosm of life.
—Piet Oudolf

For many visitors, this is an abiding image of Piet Oudolf's garden in Hummelo; however, the yew hedges are now gone, as the evolution of the garden continues. Gardens and trends may be influenced by the past, as Oudolf is, but ideas on perennial planting have been developed, modified, and progressed, so that now, under his enlightened guidance, we have learned to see the beauty of plants even when they are dead. This is gardening for the 21st century—low maintenance, chemical free, and with wildlife in mind, yet always finished with exquisite beauty and the glory imbued by nature's finesse.

This border in the Karl Foerster Garden, Potsdam-Bornim, Brandenburg, Germany, illustrates a planting in the naturalistic style of which Foerster (1874–1970) was an early exponent. Foerster, who had a lifelong passion for perennial plants, revived his parents' plants nursery in Berlin. Selecting only the best for beauty and resilience, he bred over 370 varieties, notably *Calamagrostis × acutiflora* 'Karl Foester', a form of feather reed grass, publishing his first catalog in 1907. His ideas on naturalistic style have influenced gardeners in subsequent generations and his impact remains until this day.

Jeremy Francis

b. 1951

Australia

Common beech
Fagus sylvatica

At its finest as a specimen, yet useful when trimmed as an attractive hedge, the young foliage is fresh, vibrant green; its fall color is rich copper.

For 20 years Jeremy Francis worked the family farm, 62 miles north of Perth, in the wheat belt of Australia, creating a small garden and importing perennials from the UK as a hobby. In 1990 he sold the farm and searched Australia for two years, looking for the perfect location for a new garden, finally settling on the site of an old flower farm on a 5 acre, west-facing hillside in the Dandenongs near Melbourne. Inspired by his love of plants and art, Francis has melded features of the flower farm into Cloudehill, an Arts and Crafts garden with a contemporary flavor.

In 1917, Ted Woolrich established a 5 acre nursery on the family's 10 acre block. It soon became well known for cool temperate plants, selling the first Kurume azaleas introduced by plant collector Ernest Wilson in 1922. A little later, Ted's brother Jim started growing cut flowers and foliage for florists; they shared the land until a forest fire decimated the whole site and both businesses subsided into decline, closing down in the 1960s.

Jeremy Francis had been interested in the history of garden design for many years. His wife, Valerie, had relatives living next door to Christopher Lloyd and, after a chance meeting in 1988, Lloyd helped Francis select plants that would be suitable for him in Australia. The list of other English nurseries and gardens Francis has visited over the years reads like a horticultural "Who's Who?;" he was particularly inspired by Sissinghurst, Hidcote, and Penelope Hobhouse's Tintinhull, pondering, "Could such a garden be created in Australia?"

He found the perfect location in the Dandenong Ranges. Here, the volcanic soil is deep and fertile, rainfall is reliable, and frosts are rare. He bought Jim's old flower farm in 1992; Jim had died in the previous spring. "Amongst 30 years of weeds there were screens and hedges, rare plants, magnificent historic trees, interesting open spaces, and naturalized bulb meadows. I have never known of anyone who has had the chance to make a garden from such an ideal site," Francis said.

THE ART OF GARDENING

There are now 25 different rooms and spaces, their geometry, harmony, and style reflecting the attention to detail and proportion of the Arts and Crafts movement. Several gardens were planned and constructed along the main terrace in the first few months. Since then, there has been a project every two or three years. These have been laid out with a great deal of spontaneity as the garden extended into old plantings on the outskirts of the property. "Having time to think out solutions, then being prepared to change even while the excavator is operating, is probably the ideal approach to garden making," Francis says.

There is now a water garden with a reflecting pool, shade borders, and a copper-roofed peony pavilion with moon windows, but the two Jekyll-inspired long

double borders, warm and cool, on the main terrace are the most admired feature. The warm borders appear first, red, orange, and predominantly yellow with purple-foliaged plants, a cheerful, uplifting welcome encouraging visitors to explore the garden. Beyond, the ethereal cool border, picked out in pink, blue, cream, and silver, recedes into the distance, creating a sense of perspective. Within both are plant associations, such as deep-purple-blue *Clematis* 'Polish Spirit' complementing the mahogany red hips of *Rosa sweginzowii* and foliar focal points, notably *Pyrus salicifolia* 'Pendula'.

Foliage and texture provide a constant theme throughout. Purple-leaved acers reflect a brick path, dwarf Boston ivy trails languidly over an archway, and there is a "crinkle crankle" box hedge in the peony garden. Shrubs, topiary, and brickwork

Lily tree
Magnolia denundata

This distinctive large shrub or small tree with fragrant pure white flowers in early spring always produces an abundance of blooms. There are several garden-worthy selections with different colored petals.

provide winter structure, embellished with simple plant associations such as *Salix acutifolia* 'Blue Streak', its silvery catkins shining against the bronze winter foliage of a formal beech hedge.

Several of Jim Woolrich's bulb meadows, planted in the 1930s, have been incorporated into Cloudehill. Paths have been cut through and around them. Spring bulbs are followed by wave after wave of South African bulbs, giving off color through the long grass for eight months or more.

VIVE LA DIFFERENCE

Francis incorporates other arts into his garden. There are works by 15 local artists, magnificent pots and other ceramics, two dozen pieces of contemporary sculpture, and a theater for staging Shakespeare in summer.

A workshop with the American landscape architect Martha Schwartz in 1990 inspired the construction of a Commedia dell'arte garden. This incorporates strips of Jim's bulbs and a small troupe of full-size cutout Commedia

Chinese silver grass
Miscanthus sinensis

There are a host of attractive cultivars of this clump-forming grass, renowned for their robust habit and ornamental plumes, which provide winter interest in colder climates.

characters who stand and gesture a few inches above swathes of long grass. "They are to remind visitors of Italian Renaissance gardens and their theaters, precursors of the Arts and Crafts gardens," Francis explains.

A visit to a secondhand bookshop in Kent unearthed an anthology of poetry compiled by Vita Sackville-West and Harold Nicolson. "I was particularly taken with Nicolson's translations of Virgil, generally just one or two lines, which gave the impression of haiku." Extracts inscribed on terracotta by his sister-in-law, Trish Stewart, are placed here and there to hint at a mood. Francis also collaborates with another local artist, lettercutter Ian Marr, using fragments of texts from Homer, Virgil, Chaucer, George Herbert, even St Augustine: *Solvitur Ambulando*—"It is solved by walking," written below a tiny slate carving of the Chartres Cathedral labyrinth set in the entrance path.

Inspired by many of the "greats," and with his own added ingenuity, passion for plants and love of the arts, Francis has absorbed Jim's old flower farm into the essence of an Arts and Crafts garden. It is a place of beauty—with his own contemporary twist.

And you will pull off the fragile stalks of The sad lupin with its tangle of rattling seedpods.

— Georgics 1,75. Virgil (70–19 BC), translation by Harold Nicolson

JEREMY FRANCIS

Francis has encouraged artists from other disciplines to push their creative ideas to the limit, then incorporate the results into his garden. This complements the plants and adds another layer of interest to the garden.

→ Francis searched patiently until he found the right place to create a garden. Although most people don't think in this way, the garden you are creating now may not be your final one; circumstances or opportunity may prompt a move. When looking for new properties, if you have freedom of choice, research the area thoroughly, checking the rainfall and climate; take a spade and ask permission to dig some holes to check the soils in the garden. Alternatively, use your ingenuity to adapt to conditions.

→ The gardens at Cloudehill incorporate the structure of the nursery that existed there before. When moving house, live with your new garden and make observations for at least a year before making changes; you will then understand how it behaves through the seasons. Don't remove everything; existing trees and shrubs add a sense of maturity to a new garden.

→ Every plant has to pay its rent; plants that flower briefly or have poor foliage or form don't make Francis's list. The garden is filled with plants that would be described as "garden-worthy."

→ Francis is a voracious reader. He has read works by Vita Sackville-West, Gertrude Jekyll, and other great gardeners, which have formed and influenced his opinions. "One of the great strengths of our time lies in the quality and diversity of gardening books. Find a good bookshop, have a browse, and don't be frightened to reach into your pocket." They are your guides, your textbooks, and they broaden the mind; an invaluable way to learn.

→ Bring the other arts into your garden. You may not have room for a theater but carefully placed statuary or *objets trouvés* add extra dimensions to your garden. The only limit is your imagination.

Tree rhododendron
Rhododendron arboreum

A magnificent large shrub or small tree with flowers varying in color from white to crimson. It is better grown in climates where the early flowers are not spoiled by frosts.

This view from the terrace down to the theater captures the artistic vision of the great gardener Jeremy Francis, creator of Cloudehill. The steps are flanked by *Picea glauca* var. *albertiana* 'Conica', and dwarf white spruce (a useful alternative to conical topiary). Beyond are the billowing forms of *Acer palmatum* var. *dissectum* Dissectum Atropurpureum Group, the purple cut leaved maple, imported from the Yokohama Trading Nursery, Japan, in 1928. The maples are a blaze of red and orange in fall. A hedge of *Fagus sylvatica*, the common beech, in its copper and green, overlooks the scene, the centerpiece of which is a jar by Robert Barron of Gooseneck Pottery, who wood-fires his pieces—the only potter in Australia to do so.

Will Giles

b. 1951

United Kingdom

Chusan palm
Trachycarpus fortunei

This palm is hardy down to 5 degrees Fahrenheit. Two specimens from the earliest seed introductions still stand by the main gate at the Royal Botanic Gardens, Kew.

As a small boy, Will Giles loved gardening. It later became his life. Having studied art and travelled the world, he was inspired to create a theatrical "stage set" garden in his hometown of Norwich, Norfolk. The Exotic Garden has become his *magnum opus*. This unexpected Eden is the product of a maverick imagination, out of place and time; it is a Victorian gothic fantasy, packed with extraordinary plants, theatrical ruins, and quirky surrealist art. Most of all, it brings a touch of the exotic to a cool temperate climate.

A fading photograph of a seven-year-old clutching an *Apehlandra* and cactus is early evidence of Will Giles, exotic gardener. His father moved to Norwich from London after the Second World War, in search of the "good life." The young Giles tended the plot, too, and after much pestering, his father gave him space under an old Bramley, promising: "If you can keep it weed free for a year, I will give you more next year." Later that year Giles's grandmother, Nellie, took him to the Royal Botanic Gardens at Kew. "Opening the huge doors to the Palm House, there was a blast of heat and humidity and we stepped inside to be overshadowed by

huge leaved tropical plants," he recalls. From that day, Giles was hooked. For his tenth birthday his father built him a glasshouse, thinking gardening was a passing phase, but it never passed. The glasshouse was soon crammed with exotics and after an argument with his father over heating costs he began experimenting with plants to discover which ones could be grown outside.

Several years later, Giles studied at Norwich School of Art, specializing in photography and illustration (for many years illustrating "step-by-steps" in RHS publications). In his early twenties, he scoured bookshops, desperate for more information. George Nicholson's

The Illustrated Dictionary of Gardening, with black and white line drawings of "stove plants," was an early inspiration, and Will spent hours absorbing information on anything from palms to gingers, bananas and orchids. But most inspiring of all was *Exotic Gardening in Cool Climates*, by Myles Challis, explaining how to create a garden redolent of distant tropical climes. It made him believe his vision could become reality.

AN EXOTIC VISION

Years later, in 1982, after a diligent search, Giles bought a house in the center of a neglected 1 acre plot on a south-facing slope. It took ten years of working alone, often from dawn till dusk, to clear the ground using hand tools before a raised terrace, a vegetable plot, and herbaceous borders emerged around a central lawn.

Over time the borders expanded, the lawn was dug up, and a backbone of "exotica" such as *Trachycarpus fortunei* (Chusan palm) and *Cordyline australis* (cabbage palm), which did not need winter protection, began to appear. Sinuous paths threaded through a jungle of giant leaves and riotous colors. Surrounding trees and hedges remained, providing shelter and extending the growing season by several weeks—invaluable in early

fall, when tender plants reach their peak. Temperatures occasionally slip below 25 degrees Fahrenheit in winter, rising to over 86 degrees Fahrenheit on hot summer days. Self-taught and unrestrained by preconceptions, Will revels in challenging received wisdom. A giant mat of *Tradescantia pallida*, or spiderwort, spreading under a conifer, has been outdoors for six years, surviving temperatures as low as 23 degrees Fahrenheit with the protection of the overhanging conifer proving crucial.

With the exception of hardy plants and banana plants that are too big to be moved, all the plants are dug up and stored in a large polythene tunnel in winter and vast amounts of well-rotted organic matter are

Houseleek tree
Aeonium arboreum

This architectural succulent seems more like an alien life form than a garden plant. There are several color selections, notably 'Zwartkop', with dark purple-black leaves.

WILL GILES

An exotic garden is particularly labor-intensive in spring, when tender plants are planted, and fall, when they are lifted for storage; however, anyone can create a display if the size is tailored to available time and space and the climate is taken into consideration.

→ Exotic gardening is an opportunity to experiment with hardiness. Much depends on the climate; some plants are root hardy, others survive on well-drained soils or need protection from damp and cold. Be prepared to take risks.

→ To reduce the number of plants being lifted in fall, herbaceous plants such as cannas can be cut back and covered with a 2 foot layer of straw, then plastic sheeting weighed down at the edges. The straw acts as insulation, the plastic keeps water from the roots. Both are then removed after the last frosts in spring and the plants will regrow.

→ Maintaining plants and keeping the garden tidy means you are ready for visitors at any time. Weed constantly and remove dead leaves plus fading and dead flowers, which also extends the flowering season.

→ Grow plants in pots, making them easier to lift and overwinter in a cool glasshouse or conservatory. The pots can then be plunged in the border (buried so they cannot be seen) when replanting in spring.

Indian shot plant
Canna indica

This fast-growing tender herbaceous plant has a neat compact form and bright red flowers. It should be overwintered indoors in cool temperate climates.

→ Although flowers are the most eye-catching, colored or variegated leaves provide interest through the season and are the foundation of a constant display.

→ The timing of spring planting is crucial. At the Exotic Garden, varieties tolerating temperatures just above freezing are planted out from the third week of April; more tender "exotics" around the third week in May. Tender plants rarely recover from the shock of sudden cold snaps but grow so quickly that, even when planted late, borders will burgeon from July until the first frosts.

→ Large, leafy plants need plenty of food and water. Add copious amounts of well-rotted organic matter before planting to hold moisture; feed with blood, fish, bone and pelleted chicken manure in mid-spring; and water well.

placed into the borders before they are replanted each spring. This herculean task is completed with the help of friends. Plant placement is "free form"—sometimes Giles wanders for hours, pot tucked under one arm, deliberating, pausing, contemplating, until the ideal spot is found.

Large leaves provide the structure, among them *Tetrapanax papyrifer* 'Rex' (the "rice paper" plant), its leaves reaching three feet across, and a clump of *Musa basjoo*, or hardy banana, that has been outdoors for 27 years. The remaining space has been packed with cannas, gingers, and tender bedding plants grown from seed or bought from a local houseplant wholesaler, with more than a hundred different bromeliads in the collection. Among this eclectic mix, hardy plants such as *Persicaria virginiana* (Variegated Group) 'Painter's Palette' adopt an exotic air; even *Calystegia sepium*, bindweed, takes on the guise of its tropical counterpart, *Ipomoea*.

XEROPHYTIC GARDEN

At the top of the slope, behind the house, is the xerophytic garden, built under arc lights over three months during winter 2008. The walls are made from breeze blocks, faced with local flint. It is full of little enigmas: there's a grotto with a window looking into neighboring woodland, raised beds, and hidden steps leading to nowhere. "At 17, I went to Arizona, saw *Carnegia gigantea*, the 'Saguaro,' and was overwhelmed by the urge to hug it, a painful experience that cemented my relationship with xerophytes forever!" Stocky, spiky-leaved aloes, agaves, yuccas,

and prickly pears interplanted with annuals and alliums create an arid landscape in a flint-walled island bed. They are covered during winter, to protect them from rain or snow; they can cope with being cold but dislike being cold and damp. Their angular architecture can be studied at eye level, above, or below from the looping, undulating path threading round the garden, or from the comfort of an Italianate loggia, topped with striking blue pantiles. A circular fishpond midway up the slope feeds a curtain of water that cascades spectacularly down a flint wall dotted with ferns and begonias into a trough beside the house. At the bottom of the garden, behind the polythene tunnel, a meandering path weaves through a grove of bamboos, dominated by a giant eucalyptus.

Chief among the attractive features made from recycled materials is a tree house, supported by four telegraph poles and topped by a gold-leaf obelisk, built in an oak whose boughs project at angles through the room. It provides a perfect view over the garden; from here, it is obvious that the Exotic Garden is a stage set, where everything, even the place-ment of furniture, is art. It is a living painting that proclaims Will Giles as a gardener, artist and lover of "beauty in all its forms."

My girlfriend gave me an ultimatum, "Which do you love the most, me or the garden?" I hated having to answer such a demanding question, especially as we had been living together for such a long time, but my answer had to be the garden.

—Will Giles

This huge specimen of *Doryanthes palmeri*, the giant spear lily (bought as a 6 inch tall plant), surrounded by pots of *Hosta* 'Sum and Substance', confirm the ethos behind Will Giles' Exotic Garden. Large leaves and growth so exuberant you can barely see the house (top left) create a wall of texture, punctuated with brightly colored flowers. The tree house, built in 2002, provides a bird's eye view of this spectacular scene. Hidden behind the foliage there are raised beds filled with cacti and succulents—it is hard to believe that this is not in the tropics but in Norwich, in the east of England.

Dan Hinkley

b. 1953

United States

Soloman's seal
Polygonatum hirtum

"One of the most striking in our collection... lovely clusters of pendulous green tipped white bells in late spring, resulting in crops of blue fruit. Exceptional!" (Dan Hinkley)

Dan Hinkley grew up in northern Michigan and loved looking for plants such as orchids and trilliums in the woods as a boy. After training in horticulture then teaching at Edmonds Community College, he made his name as a plant explorer in many places including the Himalayas, China, Japan, and Vietnam. His garden and nursery at Heronswood, Kingston, WA was full of rare plants, based on his collections in the wilds of the world. In 2004 he moved to a completely different habitat, a windswept and sunbaked site on a cliff overlooking Puget Sound, where a new style of gardening was to be mastered.

Dan Hinkley moved into Heronswood, near Seattle, on September 1, 1987 and created a famous nursery and display garden based on the plants he had collected on his botanical adventures around the world. His speciality was woodland plants from China and Japan, which flourished in the modified (wetter) Mediterranean climate. It became a focal point for plant enthusiasts and the annual catalogs became a collector's item, crammed with descriptions of new plants he had found, with details of their discovery described in the entry; for example,

Mukdenia rossii: "This is my collection from Sorak-san near the N. Korean border, where it grew in rocky crevices along the river that were certainly under water during spring runoff."

The garden showcased a vast array of rare and unusual trees, shrubs, vines, and perennials in shady spots beneath conifers, a quintessentially northwest garden nestled in a rich mix of hydrangeas and woodland treasures. Among them were *Ypsilandra thibetica*, a small plant with erect feathery racemes of white emerging from rosettes of glossy green foliage, and *Arisaema amurense* var. *peninsulae*, a cobra lily with a green-and-white-striped cowl, "its leaves and flowers unfolding the manner of a large butterfly emerging from its chrysalis. I collected it in South Korea in 1993, on Ulleung Island in the Japanese Sea," Hinkley wrote.

In June 2000, to the surprise of many, he sold his garden and nursery to a large seed and plant company and moved to a new home. After a period of neglect, Heronswood was bought by the Port Gamble S'Klallam Tribe in 2012 and is now being restored and redeveloped, with Hinkley as garden director.

WINDCLIFF

Hinkley's new 5 acre garden at Windcliff, on a rocky bluff on the coast of the Kitsap Peninsula, was a completely different habitat. Bands of silt and sand run through the garden; it is wet in winter and dry in summer. The garden is exposed, sunbaked, and windy but relatively frost free. "It's hot out here even on a cloudy summer day," Hinkley says. The Seattle skyline and Mount Rainier, framed at several viewpoints, are its backdrop.

The garden structure is of waterfalls, pools, and scattered rocks, which have been carefully placed to harmonize with the rugged surrounding landscape. This naturalistic landscape is filled with plants from around the world that flourish in this habitat, an eclectic mix resulting from further collecting trips to countries such as Australia and Chile. There are grasses from the Pampas, bulbs and perennials from South Africa, such as "watsonias," *Eucomis pole-evansii*, the giant pineapple lily with rosettes of massive leaves and huge spikes of white starry flowers, and *Rhodocoma capensis*, an elegant Restio with arching stems, attractive flowers, and fine foliage that even thrives in drainage ditches in the desert of the Little Karoo. There are tough gray-leaved shrubs such as lavenders from the Maquis, eucalypts from upland forests of Australia, and exotic succulents such as agaves and aloes. Hinkley has used the hardy eucalypt *Eucalyptus neglecta* for screening. He's also experimenting with six different selections of *Embothrium coccineum*, the spidery-flowered Chilean fire bush, gathered on expeditions to Chile.

Toad lily
Trycirtis hirta

Dan Hinkley wrote in his 2000 Heronswood catalog that he was offering "our collections along the Shirakura River in C. Honsuhu, in [fall] 1997," needing 'partially shaded conditions with adequate moisture'"

BEAUTIFUL BLUES

More than 60 kinds of agapanthus weave through the landscape, in shades of pale blue to deep purple, from terrace to cliff. "I inherited a clump when we moved to Windcliff and that was the start. They add an almost springlike quality to the late-summer garden, I love the rich, dreamy quality of all that blue," says Hinkley. *Agapanthus* 'Blue Leap', a selection of Hinkley's with masses of large, rich blue flowers on sturdy stems, and *A.* 'Loch Hope' are two new cultivars that he has used because they look good with grasses, the blue sky, and the deep blue of Puget Sound.

Grasses are well represented, too, with red tussock grass, *Chionochloa rubra*, forming large clumps of copper-brown foliage, and pampas grass, including *Cortaderia fulvida*, with wandlike stems and white plumes fading to cream, and the diminutive (4 foot) upright *Cortaderia selloana* 'Pumila', topped with silky plumes that dance in the breeze.

"It's about learning what likes your site, and then bulking up on those plants," Hinkley explains.

The orange-maroon peeling bark of a large native *Arbutus menziesii*, the Madroña, rising from the edge of the bluff, inspired

Seeing plants in their rightful place makes me a better gardener and teacher.

—Dan Hinkley

Pachysandra
Pachysandra axillaris

A selection of this plant now known as 'Windcliff Fragrant' was collected by Hinkley on Long Soushan in Sichuan Province in the summer of 2006.

the colors in the house and in the garden. By the third year, larger, more mature plants were available, including *Cryptomeria japonica*, the red-barked Japanese redwood, cedar, and *Cornus alba* 'Elegantissima', with its red winter stems. "Here you can see an acre at a time," Hinkley says. "It's all about a balance of texture, color, and shape; you need to step back and not look too closely."

VEGMANIA

It may come as a surprise to many but Hinkley, known as a serious plantsman and plant hunter who spends much of his time away from home, also finds time to grow vegetables. He is an accomplished cook. "We haven't bought a vegetable in two years," says the man who pickles beetroot and makes tomato sauce and salsa from the Italian heirloom varieties he cultivates along with carrots, cabbage, cauliflower, broccoli, and greens year-round.

Hinkley continues to write, garden, plant, explore, and lecture, and has been awarded the Veitch Memorial Medal for the work that is so clearly his passion. His boyhood wonder and enthusiasm for plants and gardening sustains his continuing, much shared momentum.

DAN HINKLEY

Dan Hinkley is the consummate 21st-century plantsman. Generous with his plants and knowledge and eager to search for more, he epitomizes the age old spirit of plantsmen and gardeners who just want to share their enthusiasm with others.

→ Useful advice for planting in sandy soil is to space shrubs further apart than you would in good soil. There is more competition for moisture on a hot, free-draining site and they need to grow unimpeded by competition so there is a better chance of survival.

→ "I feel it is our duty to kill as many plants as possible." Pushing the boundaries through experimentation may result in fatalities but you eventually discover which plants survive, particularly when it comes to those of borderline hardiness.

→ When planting, use small plants. "Many were in four-inch pots. It was a challenge visually, but allowed the plants the greatest opportunity to acclimatize."

→ Before the garden was planted, windbreaks were put in place using eucalyptus and pines to reinforce the existing boundary of established trees. This creates shelter within the garden for the owners and plants and a backdrop for the garden.

→ Teach children to garden. It makes them aware of the importance of their environment and may be a future source of pleasure. "As a child I sowed orange pips in my mum's kitchen and was smitten from the first orange pip that emerged out of the pot. I am fortunate to have stumbled upon something I love so much; I have had a whole life of it."

→ Hinkley has seen plants growing in their native habitat in the places they have chosen. These can then be replicated to suit their needs, though some plants are more tolerant of a range of conditions than others.

Lily of the Nile
Agapanthus umbellatus

The glorious rounded heads of funnel-shaped flowers are held high above the foliage. Grow in moist, well-drained soil in a sunny position and protect from frost over winter.

BIBLIOGRAPHY

Research sources consulted during research include books, websites, and journals. Among them: *The Times, The Financial Times, The Daily Telegraph, The New York Times, The Wall Street Journal,* and *Los Angeles Times,* plus scientific papers including the *US National Register of Historic Places, Reef Point Bulletins,* and The Royal Horticultural Society magazine, *The Garden,* the *Journal of the American Horticultural Society, Smithsonian,* and *Arnoldia.*

Books include:

Adams, William Howard (1991) *Roberto Burle Marx: The Unnatural Art of the Garden.* Museum of Modern Art.

Bennett, Sue (2000) *Five Centuries of Women & Gardens: 1590s–1990s.* National Portrait Gallery Publications.

Berge, Pierre; Cox, Madison (1999) *Majorelle: A Moroccan Oasis (Small Books on Great Gardens).* Thames & Hudson.

Betts, Edwin M; Hatch, Peter; Perkins, Hazlehurst Bolton (2000) *Thomas Jefferson's Flower Garden at Monticello* (3rd edition). Thomas Jefferson Memorial Foundation, University of Virginia Press.

Bowe, Patrick; Sapieha, Nicolas (1995) *Gardens of Portugal.* Scala Publishers.

Brown, Jane (1995) *Beatrix: Gardening Life of Beatrix Jones Farrand, 1872–1959.* Viking.

Bryant, Geoff. Ed. (1996) *The Ultimate Book of Trees and Shrubs for New Zealand.* David Bateman.

Colquhoun, Kate (2009) *A Thing in Disguise: The Visionary Life of Joseph Paxton.* Harper Perennial.

Drury, Sally; Gapper, Francis; Gapper, Patience (1991) *Gardens of England (Blue Guides).* A&C Black.

Francis, Jeremy (2010) *Cloudehill: A Year in the Garden.* Images Publishing Group Pty. Ltd.

Gatti, Anne; Lambert, Katherine. Eds. (2010) *The Good Gardens Guide 2010–2011: The Essential Independent Guide to the 1,230 Best Gardens, Parks and Green Spaces in Britain, Ireland and the Channel Islands.* Reader's Digest.

Goode, Patrick; Jellicoe, Geoffrey & Susan, Lancaster, Michael (1986) *The Oxford Companion to Gardens.* Oxford University Press.

Hertrich, William (1988) *The Huntington Botanical Gardens, 1905–1949: Personal Recollections of William Hertrich.* Huntington Library Press.

Hobhouse, Penelope (1995) *The Country Gardener*. Frances Lincoln.

Lacey, Stephen (2011) *Gardens of the National Trust*. National Trust Books.

Le Rougetel (1986) 'Philip Miller/John Bartram Botanical Exchange'. *Garden History* Vol.14, No. 1 Spring. Garden History Society.

Leygonie, Alain (2007) *Un Jardin à Marrakech Jacques Majorelle Peintre-Jardiner 1886–1962*. Editions Michalon.

Lord, Tony (1995) *Gardening at Sissinghurst*. Frances Lincoln.

Maddy, Ursula (1990) *Waterperry: A Dream Fulfilled*. Merlin.

McConnell, Beverley (2012) *Ayrlies. My Story, My Garden*. Ayrlies Garden and Wetlands Trust.

McLean, Brenda (2009) *George Forrest: Plant Hunter*. Antique Collectors' Club Ltd.

Moore, Alasdair (2004) *La Mortola: In the Footsteps of Thomas Hanbury*. Cadogan Guides.

Pankhurst, Alex (1992) *Who Does Your Garden Grow?* Earl's Eye Publishing.

Quest-Ritson, Charles (1994) *The English Garden Abroad*. Viking.

Quest-Ritson, Charles (2009) *Ninfa: The Most Romantic Garden in the World*. Frances Lincoln.

Reinikka, Merle. A. (2008) *A History of the Orchid*. Timber Press.

Robinson, William (2010) *The Wild Garden* (A New Illustrated Edition with Photographs and Notes by Charles Nelson). The Collins Press.

Russell, Vivian (1995) *Monet's Garden: Through the Seasons at Giverny*. Frances Lincoln.

Shephard, Sue (2003) *Seeds of Fortune: A Gardening Dynasty*. Bloomsbury Publishing PLC.

Spencer-Jones, Rae. Ed. (2012) *1001: Gardens You Must See Before You Die*. Cassell.

Stern, Fredrick (1974) *A Chalk Garden* (2nd edition). Faber and Faber.

Sturdza, Greta (2008) *Le Vasterival, the Four-Season Garden: How to Create Beautiful Borders for Year-Round Interest*. Les Editions Eugen Ulmer.

Veitch, James Herbert (2006) *Hortus Veitchu* (facsimile of 1906 edition). Caradoc Doy.

Wallinger, Rosamund (2000) *Gertrude Jekyll's Lost Garden: The Restoration of an Edwardian Masterpiece*. Garden Art Press.

Walska, Ganna (1943) *Always Room at the Top*. R. R. Smith.

INDEX

CREDITS AND ACKNOWLEDGEMENTS

10 © RHS | Lindley Library
14–15 © Sean Pavone | Shutterstock
16 © RHS | Lindley Library
24 © wjarek | Shutterstock
25 © Lyubov Timofeyeva | Shutterstock
29 © RHS | Lindley Library
35 © Edwin Remsberg | Alamy
40 © Gary Rogers | The Garden Collection –
Chatsworth House
41 © Charles Hawes | GAP Photos
58 © Oleg Bakhirev | Shutterstock
59 © Photograph Andrew Lawson
64 © John Glover | Alamy
65 © John Glover | The Garden Collection –
Munstead Wood
70 © Richard Wong | Alamy
71 © gardenpics | Alamy
84–85 © Wai Chan | Shutterstock
90–91 Photograph copyright Andrea Jones |
www.gardenexposures.co.uk
92 © RHS | Lindley Library
96 © Photograph copyright Andrea Jones |
www.gardenexposures.co.uk
97 © Kelly-Mooney Photography | Corbis
100 © RHS | Lindley Library
106 © RHS | Lindley Library
114 © MMGI | Simon Meaker, Le Jardin
Majorelle, Morocco, design: Jacques Majorelle
115 © Clay Perry | The Garden Collection –
Marjorelle
120 © Bill Dewey
121 © Claire Takacs
126–127 © Jim Richardson | Getty Images
128 © RHS | Lindley Library
135 © RHS | Lindley Library
136 © MMGI | Marianne Majerus, Waterperry
Gardens, Oxfordshire
137 © J S Sira | GAP Photos
142–143 © LOOK Die Bildagentur der Fotografen
GmbH | Alamy
148–149 © Malcolm Raggett
154–155 © Fondazione Roffredo Caetani
160 © RHS | Lindley Library
168–169 Photography Copyright Jonathan Buckley
173 © RHS | Lindley Library
178–179 © Martin Hughes-Jones | The Garden
Collection – RHS Harlow Carr
180 © RHS | Lindley Library
181 © RHS | Elsie Katherine Kohnlein

184–185 © MMGI | Andrew Lawson, RHS Garden,
Wisley, design: Penelope Hobhouse
190–191 © Photo: Jerry Harpur
200 © imageBROKER | Alamy
201 © © Jerry Harpur | The Garden Collection –
Design: Piet Oudolf - Hummelo, Holland
206–207 © Claire Takacs
208–209 © Will Giles
Gardener portrait sketches © Quid publishing

All other images are in the public domain. Every effort
has been made to credit the copyright holders of the
images used in this book. We apologize for any
unintentional omissions or errors and will insert the
appropriate acknowledgment to any companies or
individuals in subsequent editions of the work.

My grateful thanks to the following:

RHS: Chris Young, Rae Spencer-Jones, Brent Elliott,
Paul Cook, Jane Cuckow. Quid Publishing: James
Evans, Lucy York and Jenni Davis.

An unexpected bonus was the irrepressible upbeat
attitude and researching skills of historian Lorie
Mastemaker in the USA. Thanks also to Sorrel
Everton, Leo Hickman and Phil McCann. To those
who read and corrected: Colvin Randall, Longwood
Gardens; Rob Jacobs, Alan and Kathy Pettitt and
Jackie Kingdom from E. A. Bowles of Myddelton
House Society; Carolyn Hanbury; Elena Zappa;
Marcus Goodwin; Rick and Toby Martin; and Christina
Blandy. To those with links to or who are the great
gardeners: Nick Hamilton, Caroline Smith, Fergus
Garrett VMM, Mike Werkmeister, John Hillier, VMH,
Mary Spiller, Glyn Jones, Bev McConnell, Robyn
Goulding and Gail Griffin. Thanks also go to Ann
Welch Jennifer Trehane, Aidan Haley, Andrew Turvey,
David and Anne Edwards, Johanna Lausen Higgins
and John Massey, friend of Princess Sturdza, Lisa E.
Pearson and Gretchen Wade at Harvard University,
Daniela Guglielmi, Fabrizio Pastor, Katharine Le
Quesne and Kate Cackett. Thanks, too, to those who
provided quotes, notably Louise Godfrey.

Grateful thanks to the following for their support; Bill
and Joy Parkin, John Vickers, Dennis and Lynne
Gibbs, Dan Wilson, Michelle Daly, Catherine and
Andrew Biggs. My thanks to my wife Gill and children,
Jessica, Henry and Chloe, and to Ralph, Hugo and
Pinch, for being so longsuffering. If anyone is omitted
it is entirely my mistake; you are included, too.